Science and Fiction

Science and Fiction – A Springer Series

This collection of entertaining and thought-provoking books will appeal equally to science buffs, scientists and science-fiction fans. It was born out of the recognition that scientific discovery and the creation of plausible fictional scenarios are often two sides of the same coin. Each relies on an understanding of the way the world works, coupled with the imaginative ability to invent new or alternative explanations—and even other worlds. Authored by practicing scientists as well as writers of hard science fiction, these books explore and exploit the borderlands between accepted science and its fictional counterpart. Uncovering mutual influences, promoting fruitful interaction, narrating and analyzing fictional scenarios, together they serve as a reaction vessel for inspired new ideas in science, technology, and beyond.

Whether fiction, fact, or forever undecidable: the Springer Series "Science and Fiction" intends to go where no one has gone before!

Its largely non-technical books take several different approaches. Journey with their authors as they

- Indulge in science speculation – describing intriguing, plausible yet unproven ideas;
- Exploit science fiction for educational purposes and as a means of promoting critical thinking;
- Explore the interplay of science and science fiction – throughout the history of the genre and looking ahead;
- Delve into related topics including, but not limited to: science as a creative process, the limits of science, interplay of literature and knowledge;
- Tell fictional short stories built around well-defined scientific ideas, with a supplement summarizing the science underlying the plot.

Readers can look forward to a broad range of topics, as intriguing as they are important. Here just a few by way of illustration:

- Time travel, superluminal travel, wormholes, teleportation
- Extraterrestrial intelligence and alien civilizations
- Artificial intelligence, planetary brains, the universe as a computer, simulated worlds
- Non-anthropocentric viewpoints
- Synthetic biology, genetic engineering, developing nanotechnologies
- Eco/infrastructure/meteorite-impact disaster scenarios
- Future scenarios, transhumanism, posthumanism, intelligence explosion
- Virtual worlds, cyberspace dramas
- Consciousness and mind manipulation

More information about this series at http://www.springer.com/series/11657

Metin Tolan • Joachim Stolze

Shaken, Not Stirred!

James Bond in the Spotlight of Physics

 Springer

Metin Tolan
Fakultät Physik
TU Dortmund
Dortmund, Nordrhein-Westfalen, Germany

Joachim Stolze
Fakultät Physik
TU Dortmund
Dortmund, Nordrhein-Westfalen, Germany

ISSN 2197-1188   ISSN 2197-1196   (electronic)
Science and Fiction
ISBN 978-3-030-40108-5   ISBN 978-3-030-40109-2   (eBook)
https://doi.org/10.1007/978-3-030-40109-2

This Springer imprint is published by the registered company Springer Nature Switzerland AG.
The registered company address is: Gewerbestrasse 11, 6330 Cham, Switzerland

Contents

1

A Quantum of Physics

Bond: *"Vodka-Martini!"*
Bartender: *"Shaken or stirred?"*
Bond: *"Do I look like I give a damn?"*
(Quote from the 21st Bond movie *Casino Royale*)

Introduction

James Bond is celebrating an anniversary. The 25th film *No Time to Die* finally makes him the most successful character in film history. When Ian Fleming published the first James Bond novel entitled *Casino Royale* in 1953, he had no idea just what kind of a character he had created. The British secret agent James Bond, whom Ian Fleming named after an ornithologist from Philadelphia, is one of the best-known film characters ever.[1] He has the habit of introducing himself with the words *"The name is Bond, James Bond"* and he enjoys a worldwide popularity that could hardly be increased. The double-O number stands for the license to kill, or as his superior M once put it,[2] in the very first Bond movie *Dr. No*:

If you carry a double-O number it means you're licensed to kill—not get killed

---

[1] Ian Fleming actually owned the book *Birds of the West Indies* by ornithologist James Bond (1900–1989). In the 20th Bond film *Die Another Day* it is alluded to directly: 007 says to the Bond girl Jinx: *"Oh, I'm just here for the birds—ornithologist."*

[2] M was played in the first eleven James Bond films by Bernhard Lee. In the 17th film *Golden Eye* M became female and was impersonated by Judi Dench, who died in the 23rd film *Skyfall* and has been replaced by Ralph Fiennes ever since.

© The Author(s) 2020
M. Tolan, J. Stolze, *Shaken, Not Stirred!*, Science and Fiction,
https://doi.org/10.1007/978-3-030-40109-2_1

Figure 1.1   For many, Sean Connery is still the most popular actor of the secret agent 007

There's no better way to put it. So far, however, it has not been clarified whether it is a coincidence that his secret number "007" coincides with Russia's international dialing code.

James Bond is regarded as the epitome of the elegant Brit who eats Beluga caviar in the most beautiful places in the world, likes to drink Dom Pérignon or Château Lafite Rothschild from 1953, and never lets a beautiful woman pass by unattended (Figure 1.1). He survives unbelievable dangers, is always impossibly fit, has nerves like wire ropes, and flaunts a fabulous general knowledge.

But what does this all-purpose heartthrob have to do with physics? Actually, one should ask the exact opposite question: Does anyone seriously believe that James Bond would still live without knowledge of physics? When 007 pursues villains at breakneck speeds or flees spectacularly from his enemies, he naturally has to follow the laws of physics like everyone else, even if the scene in question seems unrealistic. In order to exploit these laws of physics for his own purposes, he should, of course, master them. So after reading this book everyone will agree with the statement that James Bond simply has to have a profound knowledge of physics, otherwise 007 would not be among the living anymore. And not only that. More than once it has been shown that he is also able to do superhuman things in mental arithmetic. Who else can solve coupled nonlinear differential equations[3] in a few seconds while sitting on a motorcycle, as Bond—as we will see—demonstrates in the opening sequence of *GoldenEye?*

According to Ian Fleming, James Bond has blue eyes and black hair. To everyone's surprise he is a child of the Ruhr area in Germany, because he was

---

[3] Experts know what that is. Non-experts can be reassured: even experts have great respect for coupled nonlinear differential equations.

born on November 11th 1920 in Wattenscheid as the son of the Scottish engineer Andrew Bond from Glencoe and a Swiss mountaineer, Monique Delacroix from Vaud. During the first five years of his life he inhaled the air of the Ruhr area and spoke better German than English, as can be seen from the "official" Bond biography.[4] He is slim, a good sportsman, excellent pistol shooter, boxer, knife thrower, and strong smoker of the brand "Morlands". The secret agent is armed with a knife on his left forearm and a Walther PPK 7.65 millimeter pistol, which was replaced by a more modern Walther P99 in 1997. Since 2012 he has used a Walther PPK/S, 9 millimeters short. James Bond is also in the enviable position of always weighing exactly 76 kilograms at a height of 1.83 meters, no matter how old he is or which actor he is portrayed by. He has a perfect body mass index[5] of 22.7. The performers have always adapted pretty closely to these ideal values. During the shooting of the film *Casino Royale* in 2006, Daniel Craig almost had the optimal dimensions with a weight of 78 kilograms and a height of 1.82 meters. James Bond will therefore always weigh 76 kilograms in all calculations in this book. This information is used when the top agent, for example, flies through the air or when he is accelerated and the force acting on him can be determined using the formula *force = mass times acceleration.* Distances or the size of objects are each indicated in "James Bond units", compared with his body size of 1.83 meters and scaled accordingly. This often makes it possible to estimate the relevant figures quite accurately. For example, we can determine the distance of the teaspoon to M's saucer, which 007 attracts with his fantastic magnetic watch in the movie *Live and Let Die.*

Other numerical values for individual scenes in James Bond films are also well known. The data for Jaws from the films *The Spy Who Loved Me* and *Moonraker* can be determined from the corresponding values of the actor Richard Kiel: 144 kilograms at a height of 2.20 meters.[6] With this information the fall of James Bond and Jaws from an airplane at the beginning of the film *Moonraker can* be analyzed in detail.

Some data are, however, less conspicuous than for example the imposing stature of Jaws. In *GoldenEye,* James Bond and a pilotless plane crash off a cliff. Here the height of this cliff is of decisive importance. The beginning of

---

[4] To be read in: *James Bond—the Authorized Biography of 007,* by John Pearson from 1973.

[5] The body mass index (BMI) is calculated according to the formula body weight in kilograms divided by body height in meters squared. A BMI between 20 and 25 is optimal. A BMI under 20 means underweight for men, over 25 the zone of overweight begins.

[6] In contrast to the precise values for James Bond, however, the data here fluctuate somewhat. Some sources speak of only 140 kilograms, and for the body size of Jaws one finds values between 2.14 and 2.38 meters.

this scene was actually shot real, so this cliff really exists, and it has a height of 2651 meters. It is clear that this figure is extremely important for James Bond, who jumps after the plane and catches it in the air. If a calculation shows, for example, that he can only enter the plane after a drop of 5000 meters, then this is certainly interesting—but unfortunately would not have helped him. In the discussion of the scenes in this book, we therefore use all available information, such as the height of cliffs or buildings, the weights of actors and equipment, or the dimensions of space stations and missiles, as far as they are known.[7]

But if important information is not known at all, then something must be done that the average citizen would not dare to reconcile with the image of a precisely analyzing physicist: one must estimate! For example, to calculate whether Jill Masterson really died of her gold plating, we need her weight. However, it is clear that information about the weight of a lady like the actress Shirley Eaton, who played Jill Masterson in *Goldfinger* in 1964, cannot be found anywhere. Therefore, her weight must be estimated as realistically as possible. From her height and general appearance it follows that she is certainly heavier than 45 kilograms and lighter than 65 kilograms. So a value of 55 kilograms seems to be quite realistic, even if it is not quite exact. This value can then be used to calculate that the lady must have died of her gold coating after about six hours. The same fate, by the way, also befalls the unfortunate Strawberry Fields in *Quantum of Solace*. After an hour of love-making with the top agent, she lies dead in her hotel room, completely covered in crude oil.

So far there are 25 official films in which 007 has to endure the most diverse adventures.[8] The series began in 1962 with the film *Dr. No*, which was followed on a yearly basis by the films *From Russia with Love*, *Goldfinger*, and *Thunderball*. *Goldfinger* is probably the most popular James Bond film ever, and it has created some myths that are still believed today. In 1967 *You Only Live Twice* reached the cinemas, in which the Scot Sean Connery appeared as James Bond for the last time. It took another two years for 007 to return to the big screen with *On Her Majesty's Secret Service*, this time played by the Australian actor George Lazenby. But this movie wasn't as successful as its predecessors, and in 1971 Sean Connery was once again persuaded to imper-

---

[7] Here *Das große James Bond Buch* by Siegfried Tesche (Militzke-Verlag Leipzig 2006, in German) with over 600 pages is an almost inexhaustible source of information (see www.siegfriedtesche.de). Books by the German Bond expert Danny Morgenstern, like *James Bond für Besserwisser—Der Tiefe Einblick in die Welt des Geheimagenten 007,* published by Damokles-Verlag 2014, are highly recommended. On the Internet, Klaus Gericke's www.jamesbondfilme.de is the best German source on the topic of 007.

[8] This refers to films by the production company EON, which holds the licenses to film all of Ian Fleming's books. Films like *Never Say Never Again* from the year 1983 or the slapstick version of *Casino Royale* from the year 1967 are therefore not among the official films.

sonate the top agent in *Diamonds Are Forever*.[9] However, he didn't want to be committed to this character as an actor and finally stopped slipping into the role of 007. James Bond was then played in seven adventures by Englishman Roger Moore. *Live and Let Die* came to the cinemas in 1973, *The Man with the Golden Gun* in 1974, *The Spy Who Loved Me* in 1977, *Moonraker* in 1979, *For Your Eyes Only* in 1981, and *Octopussy* in 1983. Finally, Moore played the agent on a secret mission once again in 1985, in *A View to a Kill*, already slightly out of shape and only fitting Fleming's ideal measurements with some effort. In 1987, *The Living Daylights* was then the premiere of the Welshman Timothy Dalton in the leading role, followed by the film *License to Kill* two years later.[10] This followed by a long break of six years in which a new 007 was found with the Irish actor Pierce Brosnan. The very first film *GoldenEye* in 1995 was a bang, which also took into account the new world political situation after the collapse of the Eastern bloc. Now James Bond was regularly seen on the big screen again: 1997 in *Tomorrow Never Dies* and 1999 in *The World Is Not Enough*, the last film in which Desmond Llewelyn plays the legendary inventor Q, who develops the technical gimmicks for James Bond that often enable him to survive in hopeless situations.[11] At the same time, however, John Cleese is already being trained as his successor and appears in the following film *Die Another Day* from 2002 in the Q role. It took another four years—until autumn 2006—when *Casino Royale* reached the cinemas. The Bond-series was restarted by first explaining how Bond became a double-O agent. Of course, a fresh new actor was needed for this task and was found in the Englishman Daniel Craig, who also played the role in the following four movies. *Quantum of Solace* followed in 2008 as an immediate sequel to *Casino Royale*. These two films show a Bond that first has to deserve his status as a double-O agent and thus does not yet receive all the technical gimmicks of his predecessors from MI6. So these films are playing before *Dr. No,* the very first film. Q does not appear because the British secret service provides top equipment only to a top agent. Bond hasn't had his martinis shaken yet either. In short: 007 is not yet the smart daredevil we all love. He is, however, a tough fist fighter who rushes from one spectacular action scene to the next and even forgets to eat up the Bond girl in *Quantum of Solace*! But not only that, the physics also falls well short in these movies. Spectacular action scenes usually don't include so much spectacular physics. When two cars collide at high

---

[9] Rumor has it that Sean Connery was "persuaded" with money.

[10] The two films with Timothy Dalton finished at the box office among the worst of all the films in the Bond series.

[11] Desmond Llewelyn died in a car accident in 1999 at the age of 85.

speed, the result is a pile of scrap metal—a physicist can't say much more about it. In 2012 *Skyfall* came to the cinemas for the 50th film anniversary of the James Bond series. 007 was here again, the old one, and with Ben Whishaw there was also a new Q. However, this was—adapted to today's times—a very young computer nerd, whose cyber abilities the new Bond needs more than the small gadgets of earlier times. *Skyfall* became the most successful Bond film of all time, with 1.1 billion dollars collected worldwide at the box office alone.[12] In 2015, the 24th film in the series, *Spectre,* was released in cinemas and continued its success seamlessly. And at last, Bond lets his vodka martinis get shaken again, not stirred. In the 25th Bond strip *No Time to Die,* the blonde Englishman is then the top agent 007 for the fifth time.

In all his adventures so far, James Bond has ordered a vodka martini 28 times, visited 38 countries, and been told 33 times that he will die. There were 60 bond girls, 31 of them brunettes, 25 blondes, and 4 redheads. A total of 16 times we can hear women whispering *"Oh, James!"* and he has sex exactly 84 times, 19 times in hotel rooms, 2 times in a London apartment, 15 times at her place, once at someone else's, 4 times on the train, 2 times in a barn, 2 times in the woods, 2 times in a nomad tent, 2 times in hospital, 2 times in an airplane, 2 times in a seaplane, once in a submarine, once in a car, once in a shower, once in a motorized iceberg, once in a space shuttle, and 26 times in, on, or under the water.[13]

At least as meticulously as these facts, the chapters of this book will physically analyze concrete scenes from the James Bond films and evaluate them as quantitatively and in as much detail as possible. All James Bond films also derive their appeal from the fact that the viewer repeatedly asks himself the question: *"Could this perhaps work somehow after all?"* when watching spectacular stunts or technical tricks. That's why we're not trying to explain how unrealistic one or the other scene is, but we're always trying to give conditions under which the scenes could actually be realized, because James Bond is not a science fiction character. However, it should come as no surprise that these conditions are sometimes somewhat unusual.

James Bond is only as good as his opponent. The villain Hugo Drax builds a large station in space and pursues a diabolical plan: he wants to destroy mankind with the help of satellites, which contain deadly poison from orchids, and repopulate the earth with flawless people selected by him. A plan that challenges a more detailed analysis. But the best James Bond opponent is

---

[12] The top three bond films by box-office takings (inflation-adjusted, as of January 2019) are *Skyfall, Thunderball,* and *Goldfinger.*

[13] These figures are based on a very careful analysis of all relevant scenes from the first 24 films.

undoubtedly Auric Goldfinger from the film of the same name. Everybody knows him, everybody knows that Goldfinger wants to break into Fort Knox, the American gold depot, and everybody knows that a so-called "atomic device" is supposed to help him, but James Bond defuses it only "007" seconds—as the time bomb counter indicates—before the detonation. But has anyone really understood in detail what Goldfinger calls his crime, the operation Grand Slam? This book will finally uncover secrets like these and even answer the question to end all questions: Why must 007's favourite drink, the vodka martini, always be shaken and not stirred?

Since the readers of this book certainly have different physics backgrounds, the sections are always divided into three parts. First, the James Bond scene is explained in detail. Afterwards the physics behind this scene is explained, avoiding complicated formulas as far as possible. At the end of each paragraph there is a section "Details for know-it-alls", where you can find out more about the relevant physics behind the given scenes and the calculations performed.

As an excellent introduction to this book, we recommend watching the two films *Goldfinger* and *Moonraker*. Firstly, you can't enjoy these classics often enough and secondly, in this book we examine all the important details from these films. But the new movies with Daniel Craig are also well worth it, so a cinema evening with *Casino Royale*, *Quantum of Solace*, *Skyfall*, and *Spectre* as preparation would also be useful. Anyway, this book is most fun if you look at the corresponding scene from the James Bond movie before each chapter. Then you will be optimally prepared to understand the analyses. However, this is not a must: all scenes are usually so well known that most have seen them before. In addition, at the beginning of each chapter we describe the relevant excerpts in detail, so that you will always be able to imagine the situation.

Basic Course Mechanics

In order to explain Bond's daring adventures we will often need *classical mechanics*. It is based on three axioms which Isaac Newton set out in 1687 in his pioneering work *"Philosophiae Naturalis Principia Mathematica"*. With the help of these three laws, all mechanical processes, be they jumps from great heights, breathtaking chases in the air, or the launch of a rocket, can be described. They are:

(1) A body moves in a straight line and at a constant speed when no force is applied to it.

(2) The rate of change of the momentum of a body is equal to the force acting on it. The momentum is given by the product of mass and speed.

(3) When two bodies interact with each other, the force exerted by the first on the second is equal to the force exerted by the second on the first, or "action = reaction" for short.

All three axioms are obviously needed to describe the adventures of 007. Using only these three relatively simple laws, all mechanical processes in nature can be understood, as long as speeds are not too high. Only when bodies move approximately at the speed of light do Newton's axioms lose their validity, and then they must be replaced by the laws of Albert Einstein's *Special Theory of Relativity*, which he found in 1905. But this will not be needed with James Bond. Although he often moves fast, compared to the 300,000 kilometers per second of the speed of light he is not soooo fast!

Newton's first axiom is also called the law of inertia, because once a body has a certain velocity, it will maintain it as long as no force is applied. Newton's second axiom, which is sometimes confusingly also referred to as the law of inertia, is often found in a special variant, because in many cases the mass of a body does not change during its motion. Then this axiom can simply be written as:

(2′) force = mass • acceleration, where the dot indicates multiplication.

In order to calculate the force acting on a body, one only has to know its mass and the rate of change of its speed. The latter is the definition of the term "acceleration" in physics. James Bond weighs 76 kilograms. Strictly speaking, we should say that his mass is 76 kilograms. His weight, the force exerted on him by gravity, is then equal to his mass $m = 76$ kg multiplied by the acceleration due to gravity $g = 9.81$ m/s^2, i.e., 76 kg • 9.81 m/s^2 = 745.5 newtons. However, as the acceleration of gravity is constant, there is no need to distinguish between the mass and the weight of a body. That's how we'll handle things throughout this book. Moreover, the value of the acceleration due to gravity is so close to the value of ten that we will often simply set $g = 10$ m/s^2 without incurring any major error in our calculations.

The unit of measurement of the force is the "newton". 1 newton is the force needed to increase the speed of a mass of 1 kilogram by 1 meter per second every second. This is a rather unwieldy statement. Forces are therefore best compared with the weight force that a corresponding mass would exert. If, for example, one asks whether 4000 newtons is a large force, then it simply follows from Newton's 2nd axiom and $g = 10$ m/s^2 that this value corresponds to the weight force of a mass of 400 kilograms. So if James Bond's arm were to be loaded with this force, it would be as if 400 kilos or eight sacks of potatoes of 50 kilos each were hung on his arm. His arm would thus be loaded by a rather large force, as the comparison with the weight force clearly shows.

The three Newtonian axioms are laws of nature, as the physicist would put it. They therefore apply universally and everywhere, at any time and in any context. Even a James Bond can't ignore them. We'll see exactly what that means for him now.

2

The Daniel Craig Bond Movies

The way in which the new name of the new Bond actor was leaked to the press even before today's official presentation is from the point of view of a real secret agent an embarrassing catastrophe: Mummy chatted. No torture, no coercion, no, Mum was just proud of her son, the new James Bond. Or to use his words to say, "Mommy wasn't shaken. She was just stirred!"

(On 15 October 2005, Normen Odenthal, anchorman for the German TV night magazine, announced the name of the new Bond actor.)

Daniel Craig has now impersonated the famous secret agent five times. With him, the James Bond series was relaunched in 2006, as the film *Casino Royale was* intended to explain how Bond became what he is today: a cold-blooded double-O agent with the license to kill (Figure 2.1). The film also strongly oriented itself on the first Bond novel with the same title published by Ian Fleming in 1953. In contrast to all other Fleming books, this material had not yet been filmed. But since there are now more films than novels about our hero, the relationship with his creator is no longer so strong.[1] The following three Craig Bond films therefore follow their own stories, among which *Quantum of Solace* stands out. With a length of 106 minutes, it is not only the shortest, but also immediately continues the story of its predecessor. In the films *Skyfall* and *Spectre* that follow, there are also repeated references to these first two Craig Bonds, so that an overarching action is created.

---

[1] Ian Fleming wrote 12 novels and 9 short stories about James Bond. With 25 films it is therefore clear that novels can no longer be the basis of the film stories.

© The Author(s) 2020
M. Tolan, J. Stolze, *Shaken, Not Stirred!*, Science and Fiction,
https://doi.org/10.1007/978-3-030-40109-2_2

Figure 2.1 James Bond was played in the last five films by Daniel Craig. Here he follows the bomber Mollaka in the movie Casino Royale

The Physical Burdens of a Secret Agent

At the beginning of the movie *Casino Royale*, James Bond faces a new challenge. After the inconspicuous tailing of the bomber Mollaka fails, he has to risk his neck to face the fugitive criminal in Madagascar.[2] The chase leads the two opponents to a large construction site and into the embassy of the fictitious country Nambutu, where 007 can finally catch up with the fugitive. The only problem: James Bond has to do without all the accessories of his agent existence this time and pursue his opponent on foot. This turns out to be not so easy, because Mollaka succeeds with spectacular jumps and acrobatic tricks in increasing his lead over his pursuer. The question soon arises: Are these death-defying moves even physically possible? Would the human body be able to endure the stresses and strains that occur?

A large part of the chase takes place on the very busy construction site. However, to get to the scaffolding first, James Bond chooses the path via the boom of a mobile crane (see Figure 2.2). He runs up in an upright stance and at the end jumps to the railing of the scaffolding. Based on the number of floors of the building and the average floor height to be assumed, the height of the crane boom can be estimated at around 16 meters. The angle of attack of the boom is not so easy to determine. Here it must be taken into account that scenes are often distorted in perspective. In order to nevertheless deter-

---

[2] By the way, Mollaka was played by Sébastien Foucan, the founder of "Freerunning", a variant of the more famous "Parkour", with more aesthetic and acrobatic elements.

Figure 2.2 James Bond can reach a construction site by walking up the boom of a truck-mounted crane. In the picture are shown the height *h* and length *l* of the boom as well as the parallel *H* and perpendicular, or normal, *N* force-components of Bond's *weight force* F_G. The angle of attack of the boom is $\alpha \approx 40°$

mine the angle with sufficient accuracy, one chooses a setting in which the camera is positioned almost perpendicular to the running direction of James Bond. This makes it possible to determine the angle of attack as about 40 degrees. Now it is easy to calculate that 007 has to cover a distance of 25 meters. However, there is another way to determine the running distance. James Bond takes twelve seconds to walk up the crane and makes about 3.5 steps per second with a span of about 60 centimeters. A short calculation also gives a length of 25 meters for the crane boom. The coefficient of static friction between the shoes of our top agent and the surface of the crane now becomes an interesting subject for physical consideration. This describes the strength of the mechanical adhesion between his shoes and the surface.[3] In general, adhesion on an inclined plane depends on the coefficient of static friction and the angle of attack. The steeper the angle, the greater the coefficient of static friction must be. This is illustrated in Figure 2.3.

The minimum static friction coefficient required to run on the slope can therefore be easily calculated. It turns out that this must fall in the range of numerical values close to those of car tires on asphalt. The secret agent's shoes, which were certainly developed by Q's department, naturally fulfilled this necessary condition and would have enabled him to pursue Mollaka via the

---

[3] The minimum static friction coefficient required is obtained from the quotient of the parallel and perpendicular, or normal components of the weight force, as shown in Figure 2.1.

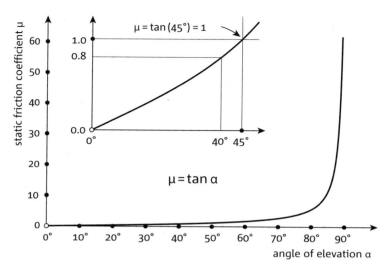

Figure 2.3 Minimum static friction coefficient μ as a function of the angle of elevation α of the crane boom. The inset shows just the angle range up to 45 degrees. For angles close to 90 degrees (i.e., a vertical wall) the coefficient of static friction increases to infinity. Most static friction coefficients lie between 0 and 1, so that inclines of a maximum of 45 degrees can be climbed upright. Typical values are: wood on stone 0.7; leather on metal 0.4; skis on snow 0.2, and car tires on asphalt 0.8

crane's boom without further problems. James Bond's analytical understanding, however, goes far beyond these rather simple calculations. He quickly realizes that the workers on the construction site have glued three sheets of commercially available roofing felt to the surface of the boom, probably for maintenance reasons. This bitumen-soaked paperboard, mixed with coarse-grained sand, is normally used as a moisture barrier in roof trusses. Due to its composition, the coefficient of static friction increases significantly and makes running on the crane a rather simple exercise for a top agent. In the film scene this roofing felt to increase the static friction on the surface of the crane boom is clearly visible.

In the next scene, first Mollaka and shortly afterwards James Bond climb up a vertical double T-beam of almost three meters in length with breathtaking speed and ease.[4] At first, this kind of locomotion might seem difficult, if not impossible, to carry out. One tends to believe that other ways of overcoming the obstacle would be more realistic. James Bond, on the other hand, is apparently immediately aware that this climbing variant is merely a special case of running on an inclined plane, which he has already mastered perfectly—as just seen.

---

[4] A T-beam is a steel beam that has the shape of the letter T in its profile. In the film you can see a double T-beam; it has the shape of a tilted H.

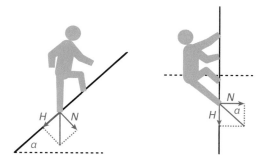

Figure 2.4 Comparison between walking on an inclined plane and climbing on a double T-beam. In one case the angle α is the angle of attack of the plane, in the other it describes the angle between the supporting leg and the direction perpendicular to the surface. In general, α describes the angle between the effective force and its normal component N, directed perpendicular to the surface. The force parallel to each surface is marked by H

A vertical beam thus corresponds to an angle of attack of 90 degrees, which makes an infinitely large coefficient of static friction necessary (see Figure 2.3). However, in contrast to walking on an inclined plane, a holding force is now also exerted. The angle of interest is therefore not the angle between the accessible surface and the earth's surface, as could be deduced from the above consideration of the physics of the inclined plane, but the angle between the direction of the force acting during climbing and the direction perpendicular to the surface. If one simplifies the situation, then for the static case, i.e., simply holding on to the beam, the right-hand picture in Figure 2.4 results.

If the static case is clear, then the moving, dynamic case is no longer a problem, because each step is only a transition between two such static cases. For example, Mollaka uses his left arm and right leg for the transition, while the other two extremities remain static. Once the two moving body parts have reached their new positions, they are used for stabilization in the next step. James Bond does not seem to be satisfied with this climbing method, which resembles the way animals would clamber up. Instead, he chooses the parallel method, which requires much more skill, in which the extremities of one side always maintain the statics, while the other side is used for locomotion. With this technique, additional effort is required to prevent the body from tilting sideways. However, both types of locomotion are possible for very highly-trained individuals.

Mollaka sees his only chance in fleeing upwards. The roof of the building is quickly reached and he has to switch to one of the two giant construction cranes. After a thrilling fight in which James Bond has to muster all his strength to avoid falling into the depths, Mollaka has no choice but to go back down the same way. In the best manner possible, he succeeds in doing this in two daring jumps. First he jumps onto the jib of the neighbouring crane, then onto the roof of a building under construction. James Bond follows close behind.

Based on the top agent's height of 1.83 meters, we can estimate the approximate heights of the jumps. Another possibility is the fall time of the two individuals. Both Mollaka and James Bond are subject to the same gravity that accelerates them in free fall independently of their mass.[5] This acceleration, which acts over the fall time, can easily be converted into the desired height. The fall time is 1.1 seconds, as a detailed evaluation of the film scene shows. James Bond therefore collides with the second crane at a final speed of almost 40 kilometers per hour. From this, we can calculate that the two crane jibs are 6.1 meters apart in the vertical direction.

The kinetic energy of 4500 joules resulting from this death-defying leap is equivalent to a good kilocalorie.[6] This is the calorific value of only 100 milliliters, i.e., five shots, of Cola light. For comparison, hamburgers from fast food chains have calorific values of about 500 kilocalories. It can rightly be said that a comparatively small amount of energy is released during the jump. This is not surprising, because the energy released by this jump is equal to the energy one would need to climb 6.1 meters.[7] Everyone would agree, of course, that not too much energy is burned when overcoming the height from the ground floor to the second floor.[8] Therefore the small amount of energy in the jump is quite understandable. At the same time, however, it is clear that we would be reluctant to get acquainted with a massive crane arm moving towards us at 40 kilometers per hour.

The decisive factor in this process is therefore not the total energy of the jump, but the extremely short period of time in which the body has to absorb this energy. Due to the short time span of deceleration after the jump, very large forces act. This follows from the fact that the speed of the falling body is decelerated to zero in a short time, and this corresponds to a very large acceleration.[9] Such a large acceleration results in a large force acting on the falling

---

[5] Due to the relatively short drop distances, air resistance can be neglected. Since there are cuts in the film, the exact fall time can often not simply be determined with the stopwatch, but requires a careful single frame analysis of the respective film scene.

[6] Kinetic energy is the energy due to the movement of a body.

[7] Here we deal with the energy due to the height in which a body is placed. It is called potential energy. Kinetic and potential energy together are called mechanical energy.

[8] The energy in food, i.e. its calorific value, therefore corresponds to a relatively large mechanical energy. That is why it is so easy to put on many kilograms, but so difficult to get rid of them.

[9] An acceleration is any change in speed, so physically speaking a deceleration is also an acceleration. Braking is sometimes referred to as negative acceleration.

body due to Newton's second law *force* = *mass* • *acceleration*. James Bond's only way to compensate for the forces that occur when landing from great heights is to maximize the time to slow down by lowering his center of gravity as far as he can. So when he hits the ground, he must kneel low. Due to the body structure, it can be assumed that the resulting braking distance is about half a meter (see Figure 2.5). A calculation shows that the complete deceleration process then takes place in less than a tenth of a second. Forces of almost 10,000 newtons occur. This corresponds to a weight of about one ton.

Evolution has made the human body relatively robust. This makes it possible to manage everyday life with all its dangers and efforts without injury, but also, for example, to achieve top performance in climbing, skiing, parachuting, and other extreme sports. Very high loads can therefore be withstood over short periods of time. What does this look like in the present case of the jump from a height of 6.1 meters? Physicists from the Massachusetts Institute of Technology in Cambridge (Boston, USA) have tackled this problem and found out what stresses the human tibia can withstand. Of course, this is only a single bone, to which the following argumentation applies as an example. The tibia with a smallest cross-section of about 3.2 square centimeters—approximately the area of a 50 cent coin—can withstand a compression pressure corresponding to 1600 times the air pressure without being damaged. On such a small area, this means that a maximum force of 50,000 newtons can be endured before the shin breaks. This again corresponds to a weight of five tons, which a tibia bone should be able to withstand at least for a short time. As a reminder, Bond's 10,000 newtons are thus far below the total of 100,000 newtons, the upper limit for two shinbones. So a shin can survive a

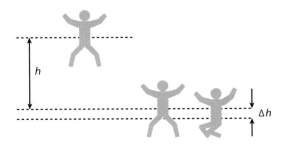

Figure 2.5 Sketch of the deceleration process at the end of a jump from a height h = 6.1 m. The center of gravity of the body must be slowed down to zero speed over the distance Δh = 50 cm in a very short time. This corresponds to a large acceleration and thus a large force acting on the body. The process can also be explained with the conservation of energy: the potential energy of the jump height h *is* added to the potential energy of the shift of the center of gravity Δh during the bending of the legs. At the same time, this total energy must be absorbed when the body buckles

jump from a height of about six meters without much damage. It should be clear that there are parts of the body that are far more sensitive than the shin, which are already severely damaged by a load of one tonne. Therefore a jump from a height of six meters will not generally be without consequence.

Mollaka won't give up. After a short hesitation he also jumps from the second crane onto the roof of a building below. This is a free fall from a height of eleven meters. Since the braking distance is still limited to half a meter by his body size, a more sophisticated technique is required to survive this jump without injuries. If Bond wants to counter the occurring forces again only by bending his knees and thus lowering his center of gravity, then he will have to cope with forces of at least 17,100 newtons for roughly 0.07 seconds,[10] and this corresponds to a weight force of 1.7 tons—that is approximately the weight of James Bond's Aston Martin. Even an agent in Her Majesty's Secret Service will not be able to do this without much pain, even if his tibia are not yet damaged.

If you watch the scene in the film carefully, however, Bond's additional tricks come to light to soften the impact. His numerous missions as a parachutist behind the enemy lines and his training in the ranks of the special units of MI6 have obviously helped him. There 007 learned that the rolling of the body distributes the resulting forces over a larger area and thus effectively reduces the pressure on individual parts of the body. This process can also be understood by means of calculations. If one compares the area of the sole of the foot, roughly 500 square centimeters, which absorbs the force in the first jump and transfers it to the ground, with the area of the back, about 5000 square centimeters, which does so in the second jump, and observes that pressure is nothing other than force per unit area, then a tenfold pressure reduction on the entire body results. Unfortunately the platform on which James Bond lands is a bit too short, and he makes the involuntary acquaintance of a metal sheet underneath.

So if jumps from great heights should be relatively easy to bear, why do we experience exactly the opposite in everyday life? Jumps from a height of six to eleven meters onto a solid surface cause at least bone fractures. The decisive factor here again is time. If we compare the time of 0.07 seconds in which James Bond has to initiate the rolling process with common processes from everyday life, then it is difficult to find a similarly short time span. Shutter speeds of SLR cameras usually range from 1/8 to 1/500 seconds. It is hard to imagine that a process as complex as unrolling can be carried out

---

[10] It's a coincidence that it's 0.07 seconds. The same deceleration process would also take only 0.07 seconds and not 0.09 seconds in the case of the double-O agent 009.

so quickly. For this reason, the braking time must be extended for jumps from great heights. In the film, this is usually achieved by the stuntmen jumping into a large pile of cardboard boxes, which significantly increase the deceleration time and thus significantly reduce the forces and pressures acting on them.

In conclusion, we can say that walking on an inclined plane as well as climbing on a vertical double T-beam is physically harmless and can be carried out by a perfectly trained secret agent. At least the first jump from the higher crane to the lower crane is possible, although not harmless and certainly not painless. In order to survive the second jump undamaged, 007 must not only be ultra-fit, but also have developed a perfect rolling technique.

Details for Know-It-Alls

To walk on an inclined plane (see Figure 2.2), the following more detailed considerations apply. The weight force of the runner is calculated as the product of his mass m and the acceleration due to gravity $g = 9.81$ m/s^2. Since the splitting of the weight force into a parallel component (the force H down the slope) and a perpendicular component (the normal force N) produces a triangle of forces (see Figures 2.2 and 2.4), these can be described using the product of the weight force with the angular functions sine and cosine:

$$H = m \cdot g \cdot sin\alpha \text{ and } N = m \cdot g \cdot cos\alpha.$$

The static friction coefficient μ is the quotient of the two force components H and N. Since the mass is contained in both components, it cancels out and the static friction coefficient is independent of the body weight. The quotient of the two angle functions sine and cosine is the tangent of the angle:

$$\mu = \tan\alpha.$$

This is sketched in Figure 2.3. For an angle of $\alpha = 40°$ the result is $\mu = 0.84$, i.e., the value for a car tire on asphalt. The static friction coefficient is a material constant and cannot be calculated using simple methods, but must be determined experimentally for each combination of two surfaces.

The following considerations apply to vertical climbing up the double T-beam. As can be seen in Figure 2.4, the observations made on the inclined plane can in principle be adopted. However, geometric variables such as the length of the legs, arms, and upper body must also be taken into account. In contrast to walking on an inclined plane, the movement now also depends on the body measurements of the climber, because they determine the maximum angles occurring in the force diagrams and thus the minimum coefficients of static friction required.

Landing after a jump from a great height is painful, that is clear. The potential energy of a body results from the product of its mass m, the acceleration due to gravity g, and the relative height h of its center of gravity. There is no absolute potential energy, but it is always given relative to a reference height. From the definition of the potential energy it follows that it increases proportionally with the height of fall. The kinetic energy is proportional to the product of the mass m and the square of the velocity v of the moving body: $E = \frac{1}{2} \cdot m \cdot v^2$. In order to determine the final velocity after a jump, it is assumed that one energy completely transforms into the other. This means that the potential energy before the jump is completely converted into the kinetic energy on impact. Thus, by equating the two energies, the square of the final velocity can simply be calculated from $v^2 = 2 \cdot g \cdot h$.

The calculation of the deceleration process by bending the legs is also based on energy conservation. At the moment when the feet touch the ground on impact, the center of gravity has just fallen down the height h. The center of gravity has not yet reached the ground. From this point on, a force F acts on the body from the ground over the entire braking distance Δh, which loads all parts of the body, including the shins, for example. The product of this force and the braking distance, i.e., $F \cdot \Delta h$, corresponds exactly to the sum of the potential energies from the drop height and the braking distance, since at the end of the process, all the energy must be absorbed. If this equation is solved for F, the result is the force acting on the body:

$$F = m \cdot g \cdot \left(1 + h / \Delta h\right)$$

This depends on the quotient of the take-off height h and the braking distance Δh. The higher the take-off point and the smaller the braking distance, the greater the force acting on the person's body. Since the force is the product of the mass and the acceleration and the acceleration is the change in speed divided by the deceleration time, this deceleration time can be calculated from the speed calculated above and the equation for the effective force. For example, the duration of 0.07 seconds for the deceleration process of the jump from 11 m height was calculated in this way.

Rollovers and Low Blows

Later in the film *Casino Royale*, Bond takes 100 million dollars in poker from his opponent Le Chiffre, which the latter absolutely needs to pay his creditors. Le Chiffre then kidnaps the Bond girl Vesper Lynd. The top agent follows the villain with his brand-new Aston Martin DBS and drives at high speed in the dark on a narrow country road.[11] At the last moment, he realizes that Vesper Lynd is tied up on the road in front of him. 007 whips the steering wheel around in a flash so that the car goes out of control. The Aston Martin overturns several times, and Bond loses consciousness. He later wakes up in captivity, sitting naked on a chair whose seat has been removed. Le Chiffre strikes his testicles with a knotted rope, a so-called monkey fist. Bond doesn't respond to Le Chiffre's demands despite the pain, so he tries to kill him. But just then the gangster is shot by the suddenly appearing chief villain Mr. White, and 007 and Vesper Lynd are surprisingly spared.

The first thing we want to think about is why the Aston Martin DBS can roll over at all on the open road.

When Bond pulls the steering wheel around, the car follows a bend. This creates a centrifugal force that pulls the car outwards. Everyone has certainly felt this force on his own body in a chain carousel at the funfair. At the same time, gravity, i.e., the weight of the car, ensures that it remains on the road. If the car makes a left turn, then both forces apply at the center of gravity, which is at a certain height above the ground, as shown in Figure 2.6. The torque of the centrifugal force now tries to turn the car around the right wheel, while its weight counteracts this. Simplifying, we now assume that the car overturns when the left wheel lifts off the ground. This is the case when the torque of the centrifugal force is greater than the torque of the weight force of the car. Now we have to use numbers. The Aston Martin DBS has a height of 1.28 meters and a width of 1.90 meters. The center of gravity of the car is situated at about one third of its height. From an analysis of the individual images of the scene on the DVD with their respective time offset and the known length of an Aston Martin DBS of 4.72 meters, we can calculate a speed of 80 kilometers per hour. This is the speed with which Bond drives the sports car along the narrow road.[12] From the geometry it follows that the centrifugal force over-

---

[11] The Aston Martin DBS has a 5.9 litre engine, which can develop a power of 517 hp. It is available in the basic version for 250,000 Euros.

[12] This is well below 300 km/h, the top speed of an Aston Martin DBS that accelerates from zero to 100 km/h in just 4.3 seconds. The makers of the stunt later announced that the car was supposed to have driven along the road at 80 mph. That would be 128 km/h. For the calculations, however, we use our measured value of 80 km/h as a basis.

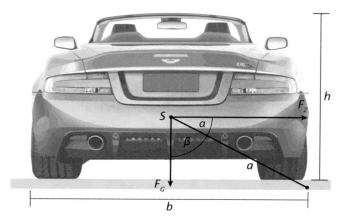

Figure 2.6 Cross-section of the Aston Martin on the road. If the car overturns, then a tire, here the right tire, is the pivot point of a rotation movement. This rotation is caused by the centrifugal force F_z, which occurs when driving along a curve. At the same time, the weight force of the car F_G counteracts this rotation. Both forces apply at the center of gravity S of the car, which has width b and height h

turns the car when the radius of the curve is less than 22 meters. Could it be that small?

Despite his whipping round the steering wheel the radius of the curve that Bond follows is certainly not much less than 100 meters. This would be about the radius of a motorway exit and thus quite tight. In the film it looks as if the radius is rather larger, i.e., more in the range of 200 meters. We can estimate this if we assume that 007 sees Vesper Lynd lying on the road in front of him at a distance of about 20 meters and passes at about one meter from her.[13] Then these values result in a radius of exactly 200 meters. But that's much larger than 22 meters, and the Aston Martin DBS shouldn't even roll over in the first place. So why is it doing it anyway?

Here it is important to know that the rollover was actually assisted in this scene. The 1750 kilogram heavy car drove over a small ramp about 10 centimeters high, so that the left wheel was lifted a little. When he was on the ramp, the stuntman used an air gun to shoot a bolt down on the left side of the car to generate additional torque and increase the centrifugal force. Without a ramp, an additional force of 36,000 newtons would be required, which corresponds to the weight force of a mass of 3.6 tons, to make the car

---

[13] If he sees her lying there when he is still 50 meters away and passes her at just one meter, the curve will have a radius of 1250 meters!

overturn even when the radius of the curve is 200 meters.[14] The ramp clearly helps here, reducing the value to only 8700 newtons. This force can be generated, for example, when an iron bolt weighing 20 kilograms is accelerated within a tenth of a second to a speed of about 10 meters per second with a compressed air gun. So you can see that here massive tricks had to be used to produce such a spectacular scene![15]

When Bond is whipped on the chair, it becomes clear that it is actually a perfidious form of torture. The monkey fist rushes up with great speed and hits him just where a male doesn't want to be hit. But why is a whip so fast and therefore so effective, even though you don't have to use much force to swing it? Analysis shows that the end of a whip can even reach supersonic speed, which can then be perceived as a whip bang. A kind of loop is formed on the rope, which moves towards the end of the whip at a rapidly increasing speed. There it opens and at its apex reaches about twice the speed of sound. For a whip with a small mass at the end, accelerations of up to 50,000 times the acceleration due to gravity can be achieved! With a larger mass, as with a monkey fist, the acceleration decreases, but the transmitted force is not smaller because of Newton's 2nd law. So it's a miracle that the tied Bond on the chair is also able to make ironic remarks, because after a painful blow from Le Chiffre he notices: *"I've got a little itch, down there. Would you mind?"*

Details for Know-It-Alls

Now the tipping over of the Aston Martin DBS will be analysed in more detail. Both gravity and centrifugal force act on the center of gravity of the car. The car rotates around the contact point of the right tire, which is at distance a from the center of gravity, and tilts over when the torque of the centrifugal force becomes greater than that of gravity. To calculate this, we must first determine the angles α and β in Figure 2.6. From the geometry, the tangent of this angle must be $\tan\alpha = 2/3 \cdot h/b$ where b is the width of the car, h is its height, and we assume that the center of gravity is located at one third of the height. If we use the numbers for the Aston Martin DBS, we get $\alpha = 24°$. For the angle β, it follows immediately that $\beta = 90° - 24° = 66°$.

In the event of overturning, the torques of the centrifugal force F_Z and the weight force F_G must then satisfy

---

[14] If one assumes the higher speed of 128 km/h indicated by the stuntman, then one needs an additional force of 29,000 newtons, for which the ramp alone would almost suffice.

[15] By the way, something was achieved with this trick. The Aston Martin turned over exactly seven times. Stunt pilot Adam Kirley thus secured himself an entry in the Guinness Book of Records!

$$F_Z \cdot a \cdot \sin\alpha > F_G \cdot a \cdot \sin\beta$$

Using the expressions for the weight force $F_G = m \cdot g$ and the centrifugal force $F_Z = m \cdot v^2/r$, the final result for the curve radius r which must be followed in order for the car to tip over at speed v is

$$r < v^2 \cdot \tan\alpha / g$$

With the figures for the Aston Martin DBS and the measured speed of $v = 80$ km/h, we get $r < 22$ m. So if the sports car has a cornering radius of less than 22 meters, it will tip over. Since the radius is much larger, however, the centrifugal force has to be assisted by a force F. The new condition for the car to overturn is

$$\left(F + F_Z\right) \cdot a \cdot \sin\alpha > F_G \cdot a \cdot \sin\beta$$

This leads to

$$F > m \cdot g / \tan\alpha - m \cdot v^2 / r$$

and results in a force of $F = 36{,}000$ N, if the mass is taken to be the mass 1750 kg of the car plus the mass 76 kg of James Bond and we use the realistic radius value of $r = 200$ m.

With a small ramp of height $H = 10$ cm and length $L = 2.5$ m, we obtain the new situation shown in Figure 2.7. Then if the car drives up this ramp only with the left tire, it is rotated around the right wheel and experiences a change in the angular velocity[16] of

$$\Delta\omega = H \cdot v / \left(b \cdot L\right)$$

where b is again the width of the car and v is its speed. This change happens in time: $\Delta t = L/v$ which leads to a change in the angular momentum of the car. However, a change in the angular momentum corresponds to a torque,[17] but

---

[16] The angular velocity is the change in the angle of rotation of the car per unit time as it drives onto the ramp and is rotated around the right wheel.

[17] More precisely, this refers to the additional torque $M = J \cdot \Delta\omega/\Delta t$. However, b now acts as the lever arm and not a, and this can be compensated by the factor $b/(a \cdot \sin\alpha)$.

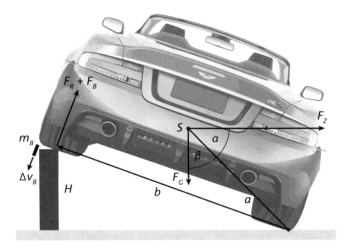

Figure 2.7   Cross-section of the Aston Martin when driving over a small ramp of height *H*. Again, the centrifugal force F_Z and the weight force of the car F_G apply at the center of gravity S and generate torques. In addition, the recoil force F_B of the bolt of mass m_B and the force of the ramp F_R provide further torques that cause the car to tip over

with the lever arm *b* and not *a* as before. This can be converted and the additional force F_R is given by the expression

$$F_R = J \cdot H \cdot v^2 / \left(a \cdot b \cdot L^2 \cdot \sin\alpha \right)$$

For the angle α we can once again set approximately $\alpha = 24°$, because the ramp is not very high.[18] J is the moment of inertia of the car in relation to a rotation around the right tire. We further assume that the car is approximately a cuboid of length *b* and height *h*. Then,[19]

$$J = 1/12 \cdot m \cdot \left(b^2 + h^2 \right) + m \cdot a^2$$

If all the numbers for Bond and the Aston Martin DBS are now used,[20] a value of $F_R = 27{,}300$ N results. The force required to tip the car over is thus reduced to 8700 N. If one now shoots a bolt with a mass of $m_B = 20$ kg, accelerated to a speed of $\Delta v_B = 10$ m/s, within $\Delta t = 0.1$ s, then a force of

---

[18] The tilting angle due to the 10 cm high ramp is only 3° for a car 1.9 m wide. This is exaggerated in the drawing.

[19] For experts, Steiner's theorem must be applied here, since the car does not rotate around an axis through the center of gravity S, but around the right tire at a distance a from the center of gravity.

[20] The length a results by applying the theorem of Pythagoras to Figure 2.6, yielding a = 1.04 m.

$F_{B,b} = m_B \cdot \Delta v_B / \Delta t = 2000$ N results because of Newton's 2nd law. However, here again, the width b of the car acts as lever arm. Therefore this must be rescaled, with the final result

$$F_B = F_{B,b} \cdot b / (a \cdot \sin \alpha)$$

Using the relevant numbers, this implies that $F_B = 9000$ N. So this is pretty much the 8700 N needed to tip the car over.

If Bond sees the lady lying in front of him when he is still at a distance approx. $d = 20$ m and passes her at a distance of $\varepsilon = 1$ m, then a curve radius of $r = d^2 / 2\varepsilon = 200$ m results from geometric considerations.

To conclude these *details for the know-it-all*, we briefly explain why the end of a whip gets to move so fast. One stroke of the whip creates a kind of loop in it, and this moves along to the end due to the centrifugal force. The strap moves away from the handle. The remaining string above the loop becomes smaller and smaller. The mass m_R *of* this remaining part of the string is proportional to its length, which decreases with increasing distance of the loop from the grip. Since the kinetic energy $E = \frac{1}{2} \cdot m_R \cdot v^2$ of the whip must be constant due to the law of conservation of energy, because after the stroke no more forces act, the velocity v must become very high. Theoretically, this mass even becomes zero at the end of the whip, so the velocity would in theory become infinitely large. In practice, however, the maximum speed is limited by friction losses, but still reaches values slightly above the speed of sound.

Bond in Free Fall with a Parachute

In the film *Quantum of Solace,* villain Dominic Greene plans to use General Medrano as head of state in Bolivia to support him in his plans to sabotage the country's water supply. Bond and Bolivian agent Camille meet at a Greene charity event and leave together to take a closer look at Greene's environmental project. They buy an old Douglas DC-3 aircraft and fly over the desert where Greene's environmental project is located. They are pursued and fired upon by a fighter plane and a helicopter. In a dramatic chase, the other plane smashes into a rock. After a collision with the helicopter, the already heavily battered DC-3 is severely damaged. But Bond can pull its nose up again and they ascend. At the highest point before the plane finally crashes, Bond and Camille can jump out of the plane with only one parachute. They are each 21 seconds in free fall and then meet in the air. This allows them to open the

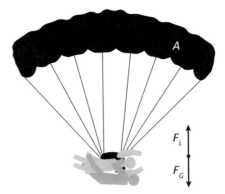

Figure 2.8 Bond and Camille hanging from a parachute with area *A*. The balance between the weight force F_G and the air friction F_L ensures a constant speed of fall

parachute only 1.5 seconds before impacting the ground. It is a kind of paraglider that unfolds fully within half a second. So they only have one second to soften the impact from such a great height. Nevertheless, they both end up in a cave without any major injuries. By chance, they discover one of the huge underground water supplies that the supposed benefactor Greene owns in the desert. That's how his perfidious plan comes to light. He has caused a drought through underground dams and now wants to sell the water supplies to the government at a high price.

We will now analyze whether Bond and Camille can really survive the fall from the plane almost unharmed (Figure 2.8).

When Bond and Camille jump out of the plane, their respective speeds initially increase. After about three seconds, they reach the terminal velocity of free fall, which results from a force equilibrium between gravity, caused by the mass of their falling bodies, and air resistance. Let us now only discuss the case of 007. For Camille, the figures would be roughly the same.[21] From his mass of 76 kilograms, his cross-sectional area of 0.8 square meters, which results from the product of his height of 1.83 meters and his shoulder width of about 45 centimeters, his c_W value of about 2, which describes the extent to which his body is streamlined,[22] and an air density of 1.2 kilograms per cubic meter, a terminal velocity of 100 kilometers per hour results, yielding a drop

---

[21] While Olga Kurylenko, who plays Camille, has a body height of 1.76 meters that comes relatively close to Bond's 1.83 meters, she has a much smaller mass than the top agent!

[22] Since Bond wants to fall as slowly as possible, he stretches out, and here we simply assume the c_W value of a flat board falling through the air. This is the absolute best case. Normally, the c_W value of a person is 0.8 and thus significantly lower. For this value, about 50% higher terminal velocities would result.

distance of about 500 meters.[23] This is realistic, because Bond had pulled the plane up shortly before bailing out.

After opening the parachute, the situation changes. Now a total weight of 139 kilograms is falling. We assume a mass of 55 kilograms for Camille and 8 kilograms for the parachute. The shape of the parachute suggests that it is a kind of small paraglider with a cross-sectional area of approximately 20 square meters and a c_W value of 3.4.[24] These values would result in a final speed of about 20 kilometers per hour, calculated in the same way as before. However, this speed cannot now be assumed to be the impact speed, because Bond and Camille only hang on the open parachute for one second and then make their sudden acquaintance with the ground. So the time is too short for them to be decelerated to this final speed. If one calculates exactly to which speed the two would be braked by the parachute after one second, the result is a value of 35 kilometers per hour. At this speed they hit the hard surface. Would they be able to stand it?

Thirty-five kilometers per hour may not sound like much. It corresponds to the speed a body would have if it fell to the ground unbraked from almost 5 meters. This shows that it doesn't have to go as smoothly as shown in the movie.[25] For example, if Bond is decelerated from this speed to zero over a distance of one meter during landing,[26] he will be subjected to a force of approximately 3500 newtons, which is about 4.5 times his weight force. This is an average over the whole body. One would have to be very well trained to survive such an impact completely undamaged. An impact at 20 kilometers per hour, on the other hand, would cause a force of only 1300 newtons, almost three times less, which would be much more tolerable.

Comparisons can also be made from road traffic. Studies have shown that a car can already cause life-threatening injuries in accidents involving pedestrians at a speed of 30 kilometers per hour. A study conducted in 2012 by Dr. Julia Seifert, Senior Physician at the Trauma Surgery and Orthopaedics Clinic

---

[23] This number results from the exact calculation and takes into account the fact that the speed at the beginning was even lower.

[24] It must also be noted here that 3.4 is quite a large c_W value for a parachute. For an open hemisphere, the actual value is 1.3–1.4, which is a much smaller value. However, for one thing we don't have a hemisphere in this case, but a narrow paraglider, and for another the parachute moves back and forth in the air because Bond and Camille move quite erratically in the air. Both factors will lead to an increase in the c_W value. With $c_W = 1.3$, approx. 50% higher values would result, and this would certainly be even less healthy for Bond …

[25] Just imagine a jump from the 5 meter tower at the swimming pool, but where someone had first drained the water from the pool!

[26] One meter is already a very generous estimate here, because Bond—always the gentleman—naturally throws himself under Camille to soften her impact. This reduces his braking distance by half, i.e., to 50 centimeters.

in Berlin, showed that in 37 accidents there were nine fatally injured pedestrians, 26 seriously injured and only two slightly injured. Spinal fractures already occurred at 33 km/h and skull fractures at 35 km/h. All the accident victims had fractures in the leg area. These figures show that you have to be pretty fit to survive a 35 kilometer per hour impact on a hard surface. We have already seen this when Bond jumped from a height of 6.1 meters onto the second crane at the beginning of the movie *Casino Royale*. Since 007 survived this and even a jump from a height of 11 meters unharmed, it is no surprise that once again not one hair is bent.

Details for Know-It-Alls

We now want to analyse in more detail the parachute jump made by Bond and his colleague Camille. To describe any kind of motion, we have to know all the forces that act on a body. According to Newton's second law, the sum of these forces is then equal to the product of the mass m and the acceleration a of the body. For a falling body, there are two forces acting on it, namely, gravity $F_G = m \cdot g$ and the opposite air resistance force $F_L = \frac{1}{2} \cdot c_W \cdot \rho \cdot A \cdot v^2$.

The latter depends on the cross-sectional area A, the density of the air ρ, the c_W value, and the square of the velocity v of the falling body. Newton's second law then implies the equation of motion:

$$m \cdot a = F_G - F_L$$

On the left-hand side we have the acceleration, which is the rate of change of the speed per unit time, and on the right-hand side we have the speed itself, we could use this equation to determine the speed as a function of the time, i.e., $v(t)$. That is what is known as a differential equation, which, even though we are know-it-alls, we will not solve here.[27]

Nevertheless we can say more. After a certain time, a body falls through the air at a constant speed. Because of Newton's first law, the total force acting on it is then zero. From the equation of motion it follows that in this case the weight force must be equal to the air resistance. If the two forces are equated, the result for the square of the speed of fall is

$$v^2 = 2 \cdot m \cdot g / \left(c_W \cdot \rho \cdot A \right)$$

---

[27] Reference is made here to the relevant literature, e.g., the textbook by Gerthsen (Springer Spektrum).

Inserting the values for James Bond returns the value $v = 27.8$ m/s $= 100$ km/h. If this speed is multiplied by the 21 seconds drop time, the distance is 584 m. The exact calculation, which takes into account the effect that Bond first has to be accelerated to 100 km/h and is therefore initially slower, yields a value of 506, i.e., about 500 meters. This shows that the acceleration time is not so important.

This is not the case, of course, when braking within one second with a parachute. Here the final speed with the numbers mentioned before in the text would be $v = 5.8$ m/s $= 21$ km/h. The calculation with the help of the exact solution of the equation of motion, on the other hand, would deliver the significantly higher value $v = 9.7$ m/s $= 35$ km/h as the speed after one second. Here our simple formula leads to a much too small value, because not enough time passes for the final speed of free fall to be reached.

We will deal with free fall in the context of air resistance in much more detail later on, when we discuss the opening sequence of the film *Moonraker*. The force exerted on a body in a 35 km/h collision with a braking length of one meter is calculated exactly in the same way as described in the first know-it-all section of this book, for someone jumping from a high scaffold.

Bond in Free Fall Without a Parachute

The film *Skyfall* begins with an incredible chase. 007 and his colleague Eve follow the assassin Patrice in Istanbul. He has stolen a hard drive containing important secret data about MI6 agents. Bond and the thief go on a wild chase by motorcycle. After the gangster has got off on a bridge and jumped on a moving train, Bond does the same and can finally confront him on the roof after further breakneck actions. They have a wild brawl on the moving train, while Eve drives to a point from where she has a good view of the train's route. She prepares a rifle to shoot Patrice when the train comes by. However, Bond and the gangster are fighting with each other, so there is no clear line of sight to shoot. At the request of her superior M, Eve finally shoots. However, she does not hit Patrice, but Bond, who then falls from the train into a river, whereupon he is washed away. Patrice can finally escape with the stolen hard drive, and MI6 declares the double-O agent dead.

First of all, we would like to investigate what forces are acting on James Bond and the gangster when they are standing on the moving train and fighting each other. Here again the air resistance is to be mentioned. It depends on their cross-sectional area, the streamlined shape of a person, and the air density and speed of the train. For the area we take as before the value 0.8 square meters and for the c_W value about 1. Bond and the villain are fighting. Therefore they don't try, as a parachutist would, to make their c_W value as big as possible, but it will rather be smaller than the value 2 attributed to a flat plate in a wind tunnel. The speed of the train is determined by an individual evaluation of the images. As Bond has a body height of 1.83 meters, there is a length against which the length of the wagons can be scaled. This analysis shows that the wagon on which Bond and Patrice are fighting has a length of about 11 meters. We then observe how long this wagon takes to pass a bush next to the track. The frame rate of the DVD results in a duration of 0.64 seconds. From this, the speed of the train can easily be calculated as 62 kilometers per hour. If we now put everything together, the force that the air resistance exerts on Bond will be 140 newtons.

This force must be compensated by the static friction caused by Bond's soles on the roof of the train. It depends on the weight of the 76 kilogram secret agent and the coefficient of static friction. This in turn depends very much on the surface quality of the two bodies rubbing against each other. If we assume that the smart top spy is wearing elegant leather shoes and that the roof of the train is made of metal, then we can set the coefficient of static friction to the value of the combination leather and metal, yielding 0.6. This results in a static friction force of approx. 450 newtons, which is significantly greater than the air resistance. Bond and the gangster therefore have no problems standing on the roof of the train travelling at 62 kilometers per hour and fighting each other. Only at speeds of 100 kilometers per hour would they get into serious trouble.

Let us now turn our attention to the case of Bond in the water. When 007 is shot by Eve, the train crosses the Gavurdere viaduct. This bridge is located on the Konya-Adana railway line, which is part of the line known as the Baghdad Railway. The imposing building is a natural stone arched bridge with eleven arches. Given its length of only 172 meters, the bridge is exceptionally high at 98 meters. Bond thus falls from a height of approximately 104 meters in total, as the height of the wagon still has to be taken into account.[28] If one now calculates the fall time for this height, then a value of 5 seconds results. This is 1.2 seconds shorter than the time that elapses in the film between Eve's shot and Bond's impact on the water. However, the fall is not shown in one piece, but through cuts one sees it from different perspectives. In the following we therefore assume the value of 5 seconds, which results from the known height of the construction. Now we can calculate the speed at which Bond hits the water. This time he falls head over heels about 100

---

[28] The height of its center of gravity is also about one meter above the height of the wagon.

meters into the water. His c_W value is about 1, the cross-sectional area only 0.14 square meters,[29] and as the density for the air we again take 1.2 kilograms per cubic meter. After 5 seconds of free fall, the result is a speed of about 140 kilometers per hour, with which 007 hits the water surface. If he hit the ground at that speed, he'd be dead instantly. But water is a liquid and therefore cushions the impact. So can he survive an impact at 140 kilometers per hour on a water surface?

To answer this, we need to calculate the force that will act on him when he dives. This force is given by the friction the body experiences when it enters the water. In principle, this is the same situation as in the case of air, except that the density is now much higher, at 1000 kilograms per cubic meter. If one then calculates the force per unit area, i.e., the pressure, which acts on Bond's skullcap during immersion, the result is a value of about 1 megapascal. That's way below 30 megapascals, which the skullcap can just about withstand without breaking. Nevertheless, one megapascal corresponds to a force of 1 million newtons per square meter. If the cross-sectional area of his head is about 300 square centimeters, this makes a short-term force of 30,000 newtons on his head. It would at least lead to a slight headache ...

Details for Know-It-Alls

We shall now justify the above figures in more detail. The drag force that acts on Bond and the gangster when they fight on the train is again

$$F_L = \tfrac{1}{2} \cdot c_W \cdot \rho \cdot A \cdot v^2$$

As in free fall, it depends on the cross-sectional area A, the density ρ of the air, the c_W value, and the square of the velocity v of the body moving through the air. The value of F_L = 140 N has been calculated using this formula.

For the adhesive friction F_H between Bond's shoes and the train roof we find

$$F_H = \mu \cdot m \cdot g$$

for the mass m of 007 and a coefficient of static friction μ between his leather soles and the metal roof. The value of F_H = 450 newtons was thus calculated.

---

[29] Here we have set a shoulder width of 45 cm again. But now Bond's thickness has to be taken as a second dimension, because he falls upside down and the relevant surface concerns the air encountered in the direction of fall. We assume a value of 30 cm here, since he is unconscious and does not fall quite straight: $0.45 \cdot 0.3 = 0.135$ m² ≈ 0.14 m².

Only when $F_L > F_H$ applies would Bond be blown away by the air resistance due to the train's motion. This is the case for speeds v such that

$$v^2 > 2 \cdot \mu \cdot m \cdot g / \left(c_W \cdot \rho \cdot A \right)$$

Putting in the numbers shows that only at train speeds of more than $v = 100$ km/h would it become difficult for our top agent to stay on his feet.

The free fall from the bridge can be treated in the same way as in the previous section. We have to take air resistance into account. However, this time Bond is far from reaching the limiting speed of $v = 96$ m/s = 345 km/h, which can be calculated from the figures given, because he does not fall for long enough. Therefore, the differential equation shown in the previous expert section must be solved exactly in order to make quantitative statements here. This leads to a velocity of impact of $v = 140$ km/h on the water.

If Bond now dives head first, then a friction force acts on him again. This time, however, it is the force that a body experiences when it moves through water. The same friction formula applies: $F_W = \frac{1}{2} \cdot c_W \cdot \rho_W \cdot A_K \cdot v^2$

The c_W value is the same as in the case of air. However, for the density ρ_W, we must use the density of water, nearly 1000 times higher, and for the area A_K, we must use the area of his head, since Bond hits the water surface with this part of his body. The pressure p on the skull is then given by the formula

$$p = F_W / A_K = \% \cdot c_W \cdot \rho_W \cdot v^2$$

Inserting the numbers finally yields a pressure on the skull of $p = 740,000$ Pa when he enters the water, which we roughly rounded up to 1 MPa.

A Helicopter, a Pistol, and a Huge Amount of Luck

The film *Spectre* follows on from its predecessor *Skyfall* and at the same time continues the story of *Casino Royale* and *Quantum of Solace*. At the end of the film, Bond is kidnapped in London and taken to the former MI6 building, which has been abandoned since the terrorist attack in the film *Skyfall*. He succeeds in taking out his kidnappers and then meets the master of all villains: Ernst Stavro Blofeld. The latter tells him that the Bond girl Madeleine Swann is somewhere in the building and that it will be blown up in three minutes. Bond is able to find Madeleine at the last moment and flee with her in a boat before the building behind them explodes. On the Thames, they track a heli-copter in which Blofeld wants to leave. Bond standing on the fast moving boat shoots at the helicopter and hits it at a crucial point. The helicopter goes into a spin and crashes onto Westminster Bridge. The head villain is just able to drag himself injured out of the wreck and crawl over the bridge. With a heavy heart, Bond does not finish Blofeld off, but after a short hesitation leaves him to M and the judiciary.

Let us now decide whether Bond's unerring shot at the helicopter was the skill of a top agent, or whether 007 was simply very lucky here. First of all, the exact distances that Bond will cover by boat and Blofeld by helicopter must be determined. This is where Google Maps can help us.

If one compares the paths followed by the boat and the helicopter against the known outline of London, it can be seen that the boat departs from the building of the MI6 upstream, as shown in the left-hand picture of Figure 2.9. It takes 40 seconds to reach the Lambeth Bridge, covering a distance of about 800 meters. This implies a speed of 72 kilometers per hour. For the helicopter, the situation is shown in the picture on the right of Figure 2.9, and we find that it covers the distance of 500 meters in 15 seconds. That implies a speed of 120 kilometers per hour.

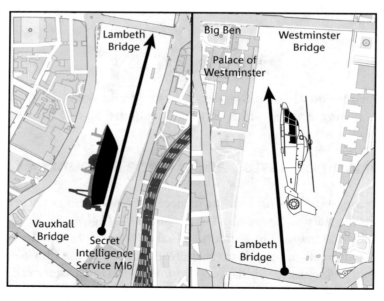

Figure 2.9 The paths followed by Bond in a motorboat (left) and Blofeld in a helicop-ter (right). Bond's path starts at the MI6 building and goes upriver. The helicopter's path starts at Lambeth Bridge

The altitude of the helicopter can be estimated by comparing it to *Big Ben*,[30] which Blofeld flies past just below the clock when he is hit by Bond's weapon. From the height of the tower at 96 meters, the flight altitude of the helicopter can be determined to be about 60 meters. Thus the geometry of the scene is known, since Bond hits Blofeld when he is right behind Lambeth Bridge. The helicopter will then be 500 meters away as we have seen before. For Bond, this means that he must fire the bullet at an angle of about 7° measured with respect to the horizontal. In the movie we can see that this is indeed the case because Bond holds his weapon almost horizontally.

The helicopter is a *Eurocopter SA 365 Dauphin 2* and has a length of 12 meters and a height of 4 meters. It is vulnerable only in the area of the engine, which is approximately one meter across. It follows from this that Bond can no longer get the helicopter from the sky if the angle of fire deviates by even 0.05°. If we also consider the fact that Bond isn't standing on solid ground and can't fix his target accurately, because he is racing across the Thames at 72 kilometers per hour, while his target is moving slightly up and down, then it quickly becomes clear that the secret agent was very, very lucky!

But it gets much better. Bond shoots with a Walter PPK/S, 9 mm short. This weapon is given to him by Q in the film *Skyfall*. A bullet has a velocity of 300 meters per second when it leaves the muzzle. It will take 1.6 seconds to reach the helicopter at a distance of 500 meters. But since the helicopter is moving at 120 kilometers per hour, it can cover a distance of about 55 meters in this time. This corresponds to almost five helicopter lengths. During this time of 1.6 seconds after it has left the muzzle of the weapon, the bullet also falls about 12 meters, which corresponds to about three helicopter heights. So Bond shouldn't even aim at the helicopter itself if he wants to hit it. Indeed, he should shoot into the air at some distance from the helicopter and then hope that the helicopter and projectile somehow coincide. This shows once again that Bond was very lucky when he brought Blofeld out of the sky at the end of the movie *Spectre*.

Details for Know-It-Alls

The top agent sees the helicopter at a height of 60 meters and about 500 meters in front of him. This is shown again schematically in Figure 2.10. With this drawing we can now calculate the shot angle α, because it is simply

$$\tan \alpha = 60 / 500 = 0.12.$$

---

[30] This refers to the Elizabeth Tower, the famous clock tower at the Palace of Westminster.

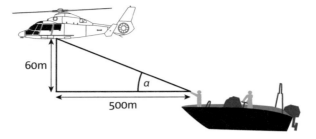

Figure 2.10 James Bond's shot at the helicopter at a distance of 500 m and a height of 60 m. Bond must adjust to the correct launch angle α

This results in an angle of α = 6.84°, which we rounded up to 7°. For the accuracy $\Delta\alpha$, with which 007 must aim to hit the helicopter, we then have $\Delta\alpha$ = 0.5/500 = 0.001.

Here it was assumed that the helicopter was vulnerable over a height of ±0.5 m, i.e., 1 m. This angular precision is given in radians and corresponds to an angular accuracy of $\Delta\alpha$ = 0.057°. Bond doesn't just have to aim precisely, he also has to have a very steady hand.

The bullet from Bond's gun is 1.6 seconds away from the helicopter's position when it was launched. The helicopter moves at a constant speed of 120 km/h in a horizontal direction. From this it can be calculated immediately that the helicopter covers a distance of 55 meters in 1.6 seconds. However, the bullet also falls a certain distance s during this time, which can be calculated using the free fall formula[31]:

$$s = \tfrac{1}{2} \cdot g \cdot t^2$$

Inserting the acceleration due to gravity g = 9.81 m/s^2 and the time t = 1.6 s, we find s = 12.5 m, i.e., the projectile falls this distance while travelling toward the target. 007 must therefore aim more than 12 meters above the helicopter so that he can hit his target.

---

[31] The air resistance does not have to be taken into account here, because the vertical speed of approx. 16 m/s achieved by the free fall of 1.6 seconds is still quite low.

3

Myths and Facts from Goldfinger

Stewardess: *"Can I do something for you, Mr. Bond?"*
  Bond: *"Eh, just a drink. A martini, shaken, not stirred!"*
  (Quote from the 3rd Bond film *Goldfinger*)

No other James Bond adventure has achieved greater popularity among fans than the 1964 film *Goldfinger*. One could almost say that this film had a lasting impact on society. Some scenes made sure that myths were formed that live on to this day. There is still controversy about whether someone whose skin is covered with a wafer-thin layer of gold would really die. This is exactly what happens in one of the most famous movie scenes ever: James Bond finds Jill Masterson on her bed completely covered in gold—and dead. So what did she die of?

The villain Auric Goldfinger wants to increase his wealth immeasurably with the operation *"Grand Slam"*. His plan is to break into Fort Knox, obviously to target America's gold reserves. But Goldfinger does not want to steal the gold. That would be problematic because of the weight of the gold ingots. But what does the bad guy hope to do then?

Auric Goldfinger acts mysteriously. In one scene he explains the operation "Grand Slam" to his fellow criminals and then leaves the room. Then we can see how an apparently poisonous gas escapes and all those left behind die instantly. The same can be observed during the attack on Fort Knox. All the soldiers stationed there fall over immediately after planes release gas. Could this creepy murder method really work that fast? What processes take place

© The Author(s) 2020
M. Tolan, J. Stolze, *Shaken, Not Stirred!*, Science and Fiction,
https://doi.org/10.1007/978-3-030-40109-2_3

during the propagation of a gas in a room, and what is the speed at which this can happen?

James Bond also caused confusion when asked whether a gun could be fired in an airplane. In one scene, the secret agent explained to Pussy Galore, who is pointing a gun at him, that her bullet will not only penetrate him, but also the plane wall, which would cause the plane to crash because of the drop in pressure. Thanks to this movie scene, most people believe that for this reason no weapon can ever be fired in an airplane.[1] So what is going on with this myth, which at first glance seems quite logical?

Goldfinger was released in September 1964 and is based on the novel of the same name by Ian Fleming.[2] The content of this film is closest to its Fleming counterpart. The novel was published five and a half years before the film in March 1959. This will become even more important when we come to the differences between the two. The Guinness Book of Records lists *Goldfinger* as one of the most successful films ever. In the first two weeks, it brought in 2.9 million US dollars—a world record, regaining almost its entire budget of three million US dollars! With 125 million dollars[3] taken at the box office alone, it achieved more than forty times its production costs. *Goldfinger* was also the first Bond film whose title song, sung by Shirley Bassey, rose to the top ten in the charts.

The lexicon of international film characterized *Goldfinger* with the words[4]:

The third Bond film is a cinematic adventure that is emphatically located beyond all credibility in the well-known, elegant mixture of science fiction, eroticism, and brutality. The political characterization of the conflicts, is only partly balanced by the unreality of the events.

Concerning the word "eroticism", we have to remember that the film was shot in the 1960s. At that time, the image of women was not too modern, as the dialogue between Bond and Felix Leiter, a CIA agent friend, clearly shows at the beginning of the film. 007 is having his back massaged in the Miami sun by Dink, a beach beauty he has just picked up.

Dink (massaging Bond's shoulders): *"Just here?"*
Bond: *"No, a little lower darling."*

[1] There are certainly many good reasons why one should not fire a weapon in an airplane, and why one should not even bring a weapon aboard an airplane. However, we do not want to discuss these here.
[2] Ian Fleming died in August 1964, shortly before the premiere of *Goldfinger*.
[3] That is equivalent to 925 million US dollars today.
[4] See the Lexikon des internationalen Films, Systhema-Verlag, Munich 1997.

Felix (just coming over): *"I thought I'd find you in good hands."*
Bond: *"Felix!"*
Felix: *"Hahaha."*
Bond: *"Felix, how are you? Dink, meet Felix Leiter."*
Dink: *"Hello!"*
Bond: *"Felix, say hello to Dink."*
Felix: *"Hi, Dink."*
Bond: *"Dink, say goodbye to Felix."*
Dink: *"Hmmm?"*
Bond: *"Man talk!"*

Bond gives Dink a loud slap on the butt on pronouncing the last word. Such a scene would no longer be possible in the 21st century, and one would never call a female film character Pussy Galore nowadays.

The film *Goldfinger* was also so successful because Ian Fleming had created Auric Goldfinger, the best possible opponent, embodied in a unique way by Gert Fröbe.[5] In the film you can clearly see this in a quotation. When the villain invites his fellow villains to his house to reveal his plans to them, he gives a short speech with the words:

> Man has climbed Mount Everest, gone to the bottom of the ocean. He's fired rockets at the Moon, split the atom, achieved miracles in every field of human endeavor—except crime!

This is a genuine announcement, which Goldfinger will redeem spectacularly, as we will see. The German translation of *Goldfinger* is, however, somewhat embarrassing. James Bond's German voice at the beginning of the film, for example, greets the inventor Q, who is about to explain the gimmicks in his new Aston Martin, with the words: *"Good morning, K!"*

Moreover, "billion" is incorrectly translated as "Billion" which means in German a million million and not "Milliarde", which is only a thousand million. This greatly influences the size of the potential gold reserves of the United States in the German version. And "outer space" does not always mean "Weltraum", i.e., the space of the universe as we will see. This is something we already actually learn in the lower level of German high school. It's really astonishing that such glaring mistakes should have been made in the translation of what is probably the most famous movie ever made.

---

[5] Note the fine tautology of Auric Goldfinger, which derives from the Latin word "Aurum" for "gold". German actors like Gert Fröbe (1913–1988) often had to serve as villains in those days, less than 20 years after the 2nd World War!

Gold: A Noble Metal

Goldfinger loves gold. This is clearly evident when his henchmen have captured Bond and tied him to a thick plate of the noble metal.

Goldfinger remarks: *"This is gold, Mr Bond. All my life, I have been in love with its colour, its brilliance, its divine heaviness. I welcome any enterprise that will increase my stock—which is considerable."*

Gold is indeed a fascinating material. It belongs to the transition metals and is in the sixth row of the periodic table. Gold has the atomic number 79 and the mass number 197, i.e., 79 electrons orbit an atomic nucleus of 79 protons and 118 neutrons. It is thus one of the elements with the most electrons and protons. Only mercury, thallium, lead and bismuth have more. Although there are also elements with significantly more electrons and protons, such as uranium or plutonium, but they are all radioactive, i.e., unstable. With a density of 19.32 grams per cubic centimeter, gold is also one of the heaviest elements. Metals such as iridium with 22.56 g/cm^3, osmium with 22.59 g/cm^3 and platinum with 21.45 g/cm^3 are even heavier, because this also depends on how densely the individual metal atoms are packed. Gold nevertheless sits in seventh place in this ranking. Microscopically, gold consists of small cubes with an edge length of 0.4 nanometers. There is a gold atom in the corners and on the surfaces of the cube. Such a structure is therefore called face-centered cubic. Gold is relatively soft and can therefore be rolled out to make the thinnest foils.[6] As with almost all metals, the melting point of 1064 °C and the boiling point of 2856 °C are relatively high. The striking colour of gold is not so easy to explain. This would require a great deal of knowledge about the microscopic structure and the energy states of the electrons in gold. And it would go well beyond the scope of this book.

The Earth's mantle consists of 0.4 millionths of a percent[7] gold, which corresponds on average to 4 grams in every 1000 tons of rock.[8] Silver, on the other hand, occurs about twenty times more frequently, with 8 millionths of a percent. By way of comparison, the mass fraction of aluminum in the Earth's crust is 7.6 percent, about 20 million times more frequent than gold. However, gold does not cost 20 million times more because of this. While one kilogram of the noble metal costs about $40,000, one kilogram of aluminum only costs a mere $2, or 20,000 times less.[9] Gold production worldwide is about 2500

---

[6] Gold leaf is only 100 nanometers thin, which is only about 250 atomic layers.

[7] This is often referred to as ppm = parts per million. 0.4 millionths of a percent is then 0.004 ppm.

[8] In gold deposits that are exploited, the proportion is usually a thousand times higher.

[9] Prices as of August 2018.

tons per year. It has increased fivefold in the last 100 years. Almost half of the gold is mined in China, Australia, Russia, Canada, and the USA. All the gold that mankind has mined so far would make a cube with an edge length of 20 meters—admittedly that would be an impressive cube that could hardly be overlooked.[10] But it wouldn't be any bigger. This shows again how rare this noble metal is.

But how did gold come into being at all? Here cosmology tells us that all elements up to iron with atomic number 26 were formed by nuclear fusion inside stars from the lightest element hydrogen. So they are the waste from the energy production of stars like our own star, the Sun. But iron is the end of the line. Heavier elements can no longer be produced by nuclear fusion in stars, since no energy would be released in the process. The gold occurring on Earth, like all elements heavier than iron, must therefore have been formed in an even larger cosmic explosion than takes place inside stars. This could, for example, have been a supernova explosion, i.e., the spectacular end of another star, a predecessor of our Sun, that has consumed all its fuel. This produces so much energy that even heavier elements than iron can be produced by nuclear fusion. But now computer simulations show that even this is probably not sufficient for the formation of gold. So what could be more powerful? Calculations carried out in 1994 showed that gold could be produced by a collision of two neutron stars. Such a collision was actually observed in August 2017 by the LIGO detectors, which are able to measure gravitational waves. These waves could be explained precisely under the assumption that two neutron stars collided at a distance of 130 million light years and caused a gigantic explosion. The reactions of the ejected matter could be observed with telescopes. They measured spectral lines which confirmed the prediction that large quantities of gold and other heavy elements were produced in such cosmic mega-catastrophes.

So gold is actually a fascinating material. But you can also view it as simply as Moneypenny in the following dialogue she has with Bond, after he has been set by M to track Goldfinger:

Bond: *"What do you know about Gold, Moneypenny?"*
Moneypenny: *"Ohhh, the only gold I know about is the kind you wear … you know, on the third finger of your left hand?"*

---

[10] One could not buy it either, because its value of 6 trillion dollars would correspond to about two times the annual gross national product of the Federal Republic of Germany!

Well, if you give a gold ring to your loved one, then you can rightly say that it must be the consequence of very distant rumblings in the universe! But now we would like to turn our attention to the film *Goldfinger*.

"Here's Looking at You, Kid!"

Even before James Bond starts his mission in the film Goldfinger, he meets the dancer Bonita, who seems to succumb to his charm. They retire to her room. In a passionate kiss, Bonita finds the gun holster Bond always wears very annoying. As a gentleman, 007 naturally takes it off. While they both turn to each other again, the hitman Capungo sneaks out from behind the closet (Figure 3.1). He has a big stick in his right hand with which the top

Figure 3.1 Fortunately, James (Sean Connery) looks deep into the eyes of his Bonita (Nadja Regin) at the beginning of the film *Goldfinger*

agent is to be overwhelmed. Bonita is obviously privy, because she is not surprised when she sees the man in her room. Fortunately, James Bond looks deep into Bonita's eyes and sees the reflection of the approaching attacker (see Figure 3.2). 007 reacts lightning fast. He pulls Bonita around, so it is she who has to take the rough blow on her head. Bond pushes her aside and begins a wild fight with the gangster, which eventually ends with Capungo in a bathtub filled with water. Even before the gangster can pull out his gun, the top agent has thrown an electrically operated fan into the tub. Capungo is killed by the electric shock, and Bond comments laconically, while Bonita still lies dazed in the room: *"Shocking! Positively shocking!"*

We now want to establish one decisive characteristic of the attacker and one of the lady through a detailed analysis of the mirror image in Figure 3.2. Let's start with the gangster. Capungo is holding the weapon in his *right* hand during his attack, so his mirror image should carry the weapon in his *left* hand, as each mirror swaps left and right. However, this is not the case. Capungos mirror image shows him clearly with the weapon in his right hand! So first of all something quite fundamental is not right with the mirror image here. There are now two possible explanations for this. The first goes like this. The film *Goldfinger* was shot in 1964. To create the reflection effect in the eye as clearly as it can be seen in the film, two film clips—an eye and a person moving towards a camera—were simply projected on top of each other on a wall and

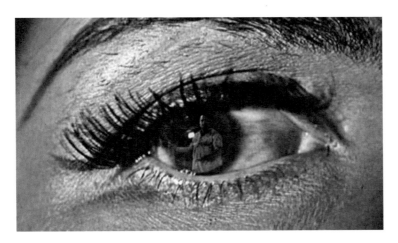

Figure 3.2 James Bond recognizes the approaching gangster Capungo as a mirror image in Bonita's eye. The image of his upper body is almost as large as the lady's iris

then filmed.[11] This fading was a common trick used in the fifties and sixties. However, by doing this, right and left are not swapped. Nowadays, this would of course be easily avoidable on the computer. One would fit the scene of the approaching attacker into the eye of Bonita with a modern video editing program and swap right and left by pushing a button, thus creating a real reflection. That works today—but of course it didn't work in 1964. At that time one had to cross fade. Explanation two for the mirror image is even easier: between the scene when you see him with the stick in his right hand and the mirror image James Bond notices in Bonita's eye, Capungo simply took the stick in his other hand. Since Bond films depict reality, this second possibility must be correct, although the first seems more plausible. But if that's the case, then Bond is dealing with a particularly dangerous opponent. Capungo is obviously an ambidextrous man who doesn't mind holding a club in his left or right hand. He can strike equally hard with both.

So this tells us that this attacker is exceptionally dangerous. Now let us analyze the mirror image in the eye a little more exactly. First of all, we must ask ourselves whether it is even possible to see the mirror image of an object in the eye of a person. This is usually not so easy to do, as anyone can test by standing in front of a mirror. Only if one illuminates objects directly in front of the eye with a strong light source can mirror images then be perceived in the eye. Figures 3.3 and 3.4 show the result of such a self-experiment, in which the scene from the film was recreated. The "attacker" can be clearly recognized in the illustrations.[12]

Figure 3.2 shows that the attacker's upper body occupies almost the entire iris of Bonita. The eye in this scene is nothing more than a spherical mirror surface. We can therefore use the laws of imaging of geometric optics to check whether the proportions of the mirror image are correct. With these laws of imaging, one can use the distance of the object to the mirror and the curvature of the mirror to calculate the size of the image on the mirror surface, thus in our case the eye surface. We assume that Capungo is about 1.80 meters tall and creeps up from about 2 meters away.[13] The radius of curvature of a human eye is about 9 millimeters, which can be estimated with the help of the curva-

---

[11] Unwanted cross-fading also sometimes occurs when photographing with very old cameras, where real films are exposed and manual forward was forgotten after taking a picture. Younger readers please ask your parents or even better your grandparents what a "film" in a camera is and what the word "develop" means in this context.

[12] The experiments show: The more tear fluid the eye develops, the better the quality of the mirror image!

[13] This assumption is reasonable, because Bond is exactly 1.83 meters tall and Capungo seems to be about the size of 007. Moreover the distance of the aggressor can be estimated quite well.

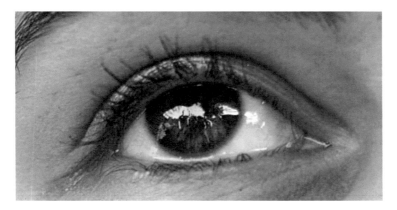

Figure 3.3 Photograph of an "attacker" at a distance of two meters. The "attacker" is clearly recognizable, but he is much smaller than Capungo in Figure 3.2

Figure 3.4 Photograph with an "attacker" at a distance of 30 centimeters. The "attacker" is now as big as in the movie scene from Goldfinger in Figure 3.2

ture of a contact lens, for example.[14] This results in 4 millimeters for the size of the attacker's image on the eye. And the image of the "attacker" in the self-experiment has exactly this size, as can be seen in Figure 3.3. Theory and experiment therefore correspond exactly—as it should be. Physics is the only science whose predictions really come out true.

Nevertheless, something fundamental cannot yet be right, because our result would mean that Capungo's upper body would appear to measure only 2 millimeters, if we assume that the size of the upper body is roughly half the body size. Since the picture of Capungos upper body fills the whole of the

---

[14] Essentially, this refers to the curvature of the cornea.

lady's iris, however, it has a size of believe it or not 1.4 centimeters in the film scene. So there's something really wrong with that.

Now this could also simply mean that we have seriously underestimated Capungo's size. This can be checked by looking at it the other way round. If we assume that the image size of 1.4 centimeters in the film scene is actually correct, then the size of the attacker can be calculated with the known curvature of the eye mirror. This results in a value of 12.5 meters! Capungo would have to be as big as a four-storey house—so that's obviously not how it's going to work![15]

But perhaps the assumed distance between Capungo and the lady's eye is also wrong. Here we assumed a distance of 2 meters, because this is the distance one would estimate from the movie scene. Therefore we repeat our considerations again, only now taking a realistic size of 1.80 meters for Capungo and the correct size of the image of 1.4 centimeters on the surface of the eye as a basis for the calculation. With this information, one can now calculate how far Capungo must be from the eye in order for his mirror image to have the same size as in the film scene. This results in a distance of 30 centimeters. This is also shown in the experiment in Figure 3.4. So, although such a large picture in the lady's eye would actually be possible, in the movie scene the distance from Capungo is much greater. If the attacker were really only 30 centimeters away, 007 would not need the mirror image to notice him, because he would have quite clearly felt his breath down his neck—so once again that doesn't work!

Now there is only one thing left that is variable. In all our considerations, only the curvature of Bonita's eye has not yet been allowed to vary. We've always assumed that the lady has a perfectly normal eye. Of course, this does not have to be so. Therefore all distances and sizes should be taken to be as realistic as possible, and we shall determine the curvature of the ocular surface. Capungo is therefore once again 1.80 meters tall, he stands at a realistic distance of 2 meters from Bonita's eye and produces an image of about 1.4 centimeters in size on the eye surface, just as can be seen in the film. Then how big must the curvature of the eye be? In fact, this gives a radius of curvature of 6.3 centimeters. In comparison, the originally assumed value of 9 millimeters was seven times smaller. The normal curvature of the eye corresponds approximately to the curvature of a two cent piece. If a circle with a radius of 6.3 centimeters is drawn next to it, as in Figure 3.5, the difference between the two curvatures is very clear.

---

[15] Theorists would certainly have no problem with this result. But since experimental physicists know that attackers are not as big as a four-storey house, we are not finished yet …

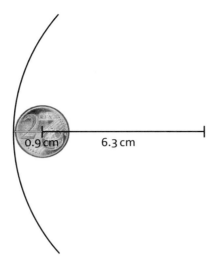

Figure 3.5   Comparison of the radius of curvature of a normal eye, which corresponds to a 2 cent coin (9 millimeters), with a radius of curvature of 6.3 centimeters, which Bonita's eye would have to have in order to explain the reflection

So it turns out that Bonita must have abnormally weakly curved eyes.[16] She would have to be completely blind, because even with the best contact lenses, she wouldn't be able to see anything anymore. So we've found out an essential feature of the lady: she is blind and only recognized James Bond by his voice or smell—at least it is absolutely clear that she wouldn't have been able to see him.

Details for Know-It-Alls

The calculation of the exact numerical values uses the laws of geometrical optics.[17] These laws of imaging apply to both lenses and mirrors. With a spherical mirror, a virtual image is created between the mirror, i.e., the eye surface, and the focal point when an object is depicted.[18] The distance between the object, i.e., the person, and the mirror is the object distance g, *and the* distance between the virtual image and the mirror is the image distance b.

---

[16] We have only analysed one eye in detail, but it would be absurd to assume that both eyes developed completely differently. Therefore, both eyes probably have about the same abnormally weak curvature.

[17] In geometrical optics, the light propagates in a straight line in the form of rays. This assumption is always justified when all objects are much larger than the wavelength of light. This is certainly the case with the scene in question.

[18] A virtual image is created by extending the light rays from the mirror surface in a backward, straight line. You can see it but you can't see it on a screen.

Furthermore, G denotes the size of the object to be imaged, B the size of the image, and f the focal length of the mirror. These quantities are shown in Figure 3.6.

Now the following equations apply, which are also called the laws of imaging:

$$B/G = -b/g$$

This equation tells us that the ratio of the image and object sizes behaves in the same way as the ratio of the respective distances from the mirror. The fact that a spherical mirror creates a virtual image is taken into account by a negative image distance $-b$. Furthermore, the lens equation applies, since it is also applicable for mirrors:

$$1/f = 1/(-b) + 1/g$$

This equation relates the distances of the object and the image to the focal length of the mirror. In addition, the radius of curvature of the spherical mirror r is twice the focal length: $2 \cdot f = -r$.

The minus sign here means that the mirror is curved outwards. With these formulas the numbers used in the text can be easily calculated.

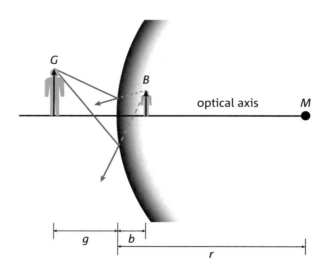

Figure 3.6 The dimensions of the spherical mirror. The object distance is indicated by g, the image distance by b, and the radius of the mirror by r. An object of size G creates a virtual image of size B, which is only created if the reflected rays are extended backwards in a straight line, as a viewer looking from the left would do

How Did the Golden Lady Die?

James Bond is commissioned by M to monitor the activities of Auric Goldfinger, suspected of large-scale gold smuggling. He gets the order to fly to Miami and settle in the same hotel as the gangster. 007 observes a young girl helping Goldfinger to cheat his opponent at a poker game. He tracks her down in her hotel room, introduces himself Bond-style by saying: *"My name is Bond—James Bond"*, and learns that the girl is Goldfinger's assistant, Jill Masterson. With his usual charm, 007 succeeds in impressing the lady in no time at all and pulling her up close to him. The two enjoy a few nice hours in the hotel room. When Bond goes into the next room to get a new bottle of champagne from the fridge, he is knocked out with a blow from the edge of a hand. Only the shadow of the perpetrator can be seen, wearing a hat. He is obviously a strong man and later turns out to be Goldfinger's henchman, Oddjob. After an unknown period of time James Bond comes round, straightens up, and staggers back into the bedroom, still slightly dazed. He hesitates and turns on the light. Then comes one of the most famous film scenes ever: on the bed Jill Masterson lies naked, covered from top to bottom with a shimmering golden layer.[19] James Bond steps up to the bed, feels her pulse, and finds she's dead. He then calls his CIA buddy, Felix Leiter:

Reception: *"Yes, Mr. Bond?"*
Bond: *"Beach 79432. Room 119."*
Leiter: *"Hello?"*
Bond: *"Hello, Felix. Get over here right away."*
Leiter: *"What's happened?"*
Bond: *"The girl's dead."*
Leiter: *"Dink?"*
Bond: *"No. Masterson, Jill Masterson. And she's covered in gold—yes, in gold!"*

The last sentence is spoken in the German translated version of this movie. The original English version differs here slightly but significantly as we shall see later. However, we will go on with the German version for the moment.

Bond is talking about the fact that Jill Masterson has been gold plated from top to toe. If this were true then this needs to be analysed in more detail. How was the gold layer applied, and can it really be real gold? The possibilities for gilding a human being are very limited. Assuming the gold is applied in mol-

---

[19] A pillow was placed so cleverly that, even in 1964, this scene did not cause any outcry because of too much sexual freedom. This pillow is surely the most correctly placed pillow in film history!

ten form, this will not go without leaving visible traces on the body, since the melting point of gold is 1064 degrees Celsius. The burns that would result would drastically change Jill Masterson's appearance. But there's no sign of disfigurement in the scene—on the contrary, Jill Masterson's body is absolutely flawless. A coating with liquid gold is also unlikely, as there are no further traces of gold anywhere in the room or on the bed. Where else could the gilding process have taken place? Taking Jill Masterson away and gold plating her somewhere else would have taken a long time and would have attracted unnecessary attention. A coating of real liquid gold can therefore be ruled out. It is undisputed, however, that such a coating would kill the lady—no question (Figure 3.7).

Another possibility for gilding something is the application of gold leaf. However, this is a very costly method. The wafer-thin gold leaves are applied individually and with great care to the object to be gilded, using a brush. This means it would take even longer and would probably not be practical for Goldfinger's purposes either.

Fortunately, the answer to the question is not that difficult: In the German version of Goldfinger, James Bond says in the scene that the lady is *"completely covered with gold"*. In the English original, however, he says literally:

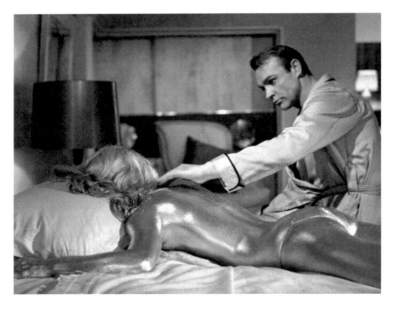

Figure 3.7 The most famous scene from James Bond movies: Jill Masterson (Shirley Eaton) lies dead on the hotel bed. Apparently, she is completely covered with a shiny golden layer. But could that really be the cause of her death?

And, she's covered in paint—gold paint.

Aha! So this is not gold, only gold paint. Colour is of course much easier to apply and is therefore the easiest way to gild Miss Masterson.

The answer to our question, however, immediately raises a new one. If Jill Masterson was painted with gold paint, what did she die of? She really is dead. At a later meeting with M, 007 tells him exactly how the lady died:

M: *"Gold? All over?"*
Bond: *"She died of skin suffocation. It's been known to happen to cabaret dancers. It's all right as long as you leave a small bare patch at the base of the spine to allow the skin to breathe."*
M: *"Someone obviously didn't."*
Bond: *"And I know who."*

So Bond claims the lady suffocated because she couldn't breathe through her skin. Many people still believe in this reasoning today.[20] It was given personally by the omniscient 007 in one of the most famous films of all time, so it must be right. But can it really be? The specialists from *"MythBusters"*, a program on the American television station *Discovery Channel,* have carried out corresponding investigations for this purpose.[21] They painted an employee with gold paint and observed the reactions of the body under medical supervision. The most important value, namely the oxygen saturation of the blood, did not experience the slightest change due to the colour layer. And the chief physician confirmed the result that one cannot suffocate by being coated with gold paint. Although there are lungless salamanders in the animal kingdom,[22] which breathe through the skin, humans have lungs and breathe in oxygen, which then passes into the blood. Only a maximum of one percent of the oxygen is absorbed through the skin. That's the way it has to be, otherwise swimming would be a deadly undertaking. A large part of the skin is also under water. So we can safely rule out Bond's explanation for Jill Masterson's death. Even the best secret agent can be wrong. But there is still another pos-

---

[20] The film set obviously also believed in this myth, as production designer Ken Adam of the film *Goldfinger* explained in the documentary *"Best ever Bond"* from 2002.
[21] *MythBusters* was an American documentary series on the *Discovery Channel.* It examined urban myths, especially those triggered by films. From 2003 to 2016, a total of 278 episodes ran. The myth of a deadly coating with gold paint was examined in the 3rd episode, which was broadcast in 2003.
[22] For example, the Eschscholtz salamander (Ensatina eschscholtzii) from the family of the tailed amphibians.

sible cause of death, which the doctor of the *"MythBusters"* also assumes to be probable: death by overheating. So why is that?

Normally, we release part of our body heat to the outside via various mechanisms and thereby regulate our body temperature. Due to the golden layer, this temperature regulation will obviously no longer take place, or only to a limited extent, and the body will gradually heat up. A body temperature of 42 degrees Celsius over a certain period of time will inevitably lead to death. So this could be how Jill Masterson's demise was brought about. However, we need to analyse this in more detail and in particular clarify how long it will take for death from overheating to occur. So we also need to know how long James Bond was unconscious before waking up and finding Jill Masterson dead.

For an initial overview, we begin by considering the case of complete insulation. It is therefore assumed that the gold layer applied completely shields the heat emitted by the body and that the body continues to heat up. A normal average person radiates a heat output of about 100 watts.[23] Since Jill Masterson is rather petite, we assume in the following only 70 watts of radiation power for her, at a body temperature of 37 degrees Celsius.[24] Now we have to determine how long it will take for her body to warm up by five degrees Celsius from the inside out. Of course her weight also plays an important role. Due to the appearance and height of the actress Shirley Eaton, we can estimate this at 55 kilograms. These 55 kilograms are thus constantly heated with an output of 70 watts. Because of the gold layer, none of this energy is released back into the environment. A calculation shows that after about 3.5 hours, the lady would die because her body would have heated up to 42 degrees Celsius. So James Bond was unconscious for at least that long.

However, our observations still contain one very great weakness: the assumption that the power radiated by the body remains completely in the body would certainly be justified if the lady were wrapped in polystyrene or another very good insulating material. Gold and gold paint, however, do not have these good insulating properties.[25] A more exact treatment is required, taking into account the heat flow through the gold layer. So what is interesting now is the question of how much heat is dissipated to the outside. Some substance-specific data about gold or gold colour are required for this. In addition, the room temperature now also becomes important, since the speed

---

[23] That's why people usually wear clothes. This quite considerable power of 100 watts should not be released unhindered into the environment.

[24] It is well known that 36 to 37 degrees Celsius are more normal. Due to the previous hours with 007, it seems realistic to assume a slightly increased body temperature for Jill Masterson.

[25] If this were the case, houses would simply have to be painted with gold paint in order to achieve optimum thermal insulation.

of heat dissipation depends on the prevailing temperature gradient, i.e., the difference between the body and room temperatures.[26] In view of the fact that the scene takes place in Miami Beach, the room temperature can be set at 28 degrees Celsius. The lady's body area must also be included in the calculation. For women like Jill Masterson, this would be about 1.6 square meters. In addition, the heat emitted depends on the thickness of the gold layer, which we assume to be 0.1 millimeters, similar to a layer of varnish. With these numbers, the amount of heat that can be dissipated to the outside through the gold layer is a gigantic 45 million watts.[27] This means that, with the assumptions made, such a large amount of heat could be dissipated to the outside that Jill Masterson's 70 watts would be of no significance at all. Her 70 watts would have gone unhindered through the gold layer, and she wouldn't heat up!

Conversely, one can now ask, just for fun, how thick a layer of gold would have to be for an output of 70 watts to lead to a noticeable change in body temperature. With the available numbers a value of 64 meters results. If Miss Masterson was to be placed in a massive gold ball with a radius of 64 meters, part of the heat radiated by the body would remain within the gold layer and gradually warm the girl. But there is no sign of a gold ball with a diameter of 128 meters in the film.[28]

If, however, the gold layer were to partially reflect the heat back inwards, like the rescue blanket in a first aid kit, then the thing could work.[29] In a gold layer, the part of the heat radiation that is absorbed is only about two percent; the rest is reflected. So 68.8 watts remain in the body and it will take a total of three hours and 36 minutes until death by overheating occurs.

Unfortunately, this argument also contains a big mistake: unlike the rescue blanket, where there is air between the body and the reflective layer, gold plating puts the body and the gold layer in direct contact. In this case no reflection can take place, so the rescue blanket effect cannot occur at all.[30] Therefore, it can't be that simple.

---

[26] Experience teaches us that hot coffee cools down faster in the snow than in the blazing summer sun.

[27] This results in the calculation for a thin layer of pure gold. Similar numbers result for a layer of gold paint. Forty-five million watts is the output generated by 45 wind turbines and is sufficient to propel three double traction high-speed trains, called ICE in Germany.

[28] Apart from the fact that we just don't have that much gold, it would also be a very expensive affair— even for a multimillionaire like Auric Goldfinger. Because with the gold price of 1281 US dollars per kilogram valid at that time, such a gold layer would cost the small fortune of around 27 trillion US dollars. That would be a lot of effort just to murder a disloyal employee.

[29] For this reason, rescue blankets should always be used with the golden side facing inwards.

[30] A perfect rescue blanket would lead to death by internal overheating just like a perfectly reflecting layer of gold. However, a blanket is never perfect and lies directly on the body in various places, so there is no risk of overheating for accident victims.

How else does the body give off heat? One of the most important and effective mechanisms is sweating. Through evaporation of sweat on the skin, heat is released and the body is cooled. On average, a person produces about half a liter of sweat per night. If one now assumes that the average person sleeps about eight hours per night, this results in a power output of 42 watts due to sweating. However, the skin's own cooling system can no longer function under a complete coating of gold paint, because the sweat glands are sealed. Forty-two of the 70 watts of power cannot therefore be emitted by Jill Masterson, and the remaining heat will lead to heating of the sealed body. About six hours will then pass before the critical body temperature of 42 degrees Celsius is reached.

That would be quite a realistic scenario and most likely the reason for Jill Masterson's tragic death. Like 007, she was beaten unconscious by Oddjob, Goldfinger's particularly strong accomplice, and painted with gold paint to tell James Bond to stay out of Goldfinger's business. Her death didn't necessarily have to be scheduled. If she had awakened faster from the blow, she would not have died of her own body heat. But our considerations now show that James Bond also received such a strong blow that he must have been lying unconscious on the floor in the next room for at least six hours. That would solve the mystery.

Details for Know-It-Alls

The basis of all calculations for the overheating of the golden lady is the direct connection between the temperature change ΔT of a body and the energy or quantity of heat Q required to achieve it:

$$Q = c_{lady} \cdot m_{lady} \cdot \Delta T.$$

The specific heat c_{lady} is a material constant and m_{lady} is the mass of Jill Masterson. The specific heat for water has a value of about 4000 J/(kg·K). This means that an amount of energy of 4000 joules is needed to increase the temperature of one kilogram of water[31] by one degree. Since the human body consists of 80% water, the heat capacity of the lady can be taken as c_{lady} = 3200 J/(kg·K). This number then means that an amount of heat equal to 3200 joules is needed to heat one kilogram of Jill Masterson by 1 °C.

---

[31] That's one liter.

This makes it easy to continue the calculation. For the power P delivered by the human body the following applies: $P = Q/t$.

The power is therefore the energy that is released into the environment per unit time. We now assume that the body releases only heat energy. If Q is the energy needed to overheat the body, then at a known output of 70 watts, we can calculate the time t required to heat Jill Masterson by 5 °C. The formula is as follows:

$$t = c_{lady} \cdot m_{lady} \cdot \Delta T / P.$$

In each case, inserting the numerical values from the text provides the times which must elapse until the occurrence of death.

The heat of evaporation Q_V of water must be taken into account during sweat formation and subsequent evaporation. For the energy E that is required to evaporate a quantity m_{sweat} of sweat, the following applies: $E = m_{sweat} \cdot Q_V$.

The evaporation heat for water is $Q_V = 2400$ J/g, i.e., an energy of 2400 joules is required to evaporate 1 gram of sweat. The evaporation rate is then

$$P = m_{sweat} \cdot Q_V / t_{sleep}.$$

Here $t_{sleep} = 8$ h is the average sleep time of a person and $m_{sweat} = 500$ g the amount of sweat produced. Inserting these numerical values, we obtain $P = 42$ watts. This power must then be used in the denominator of the above formula instead of the previous value of 70 watts, in order to calculate the actual overheating time to be $t = 6$ h.

The heat conduction through a layer of thickness d and area A is found using Fourier's law: $P = \lambda \cdot A \cdot \Delta T / d$.

Here P is the thermal output that passes through the layer for a temperature gradient of ΔT. The thermal conductivity λ is another material constant. It is 310 W/(m·K) for gold. Since the thickness of the layer is the denominator, a very thin layer of a material with good thermal conductivity naturally yields a very high thermal conduction.

A New Toy: Laser Beams

James Bond is of course not impressed by the little incident with the golden lady and sticks to Auric Goldfinger's heels. While spying on the villain's Swiss gold foundry, he meets Tilly, the sister of the murdered Jill Masterson. While the two scout around the smuggling centre, Tilly is killed by security guards, and James Bond, knocked unconscious by a car crash, falls back into the hands of Goldfinger's henchman Oddjob. When the secret agent wakes up, he lies tied up on a gold plate in a laboratory of the gold foundry. Goldfinger enters the room, and we hear the following dialogue:

Goldfinger: *"Good evening"*
Bond: *"My name is James Bond."*
Goldfinger: *"And members of your curious profession are few in number. You have been recognized. Let's say by one of your opposite numbers, who is also licensed to kill. That interesting car of yours! I, too, have a new toy, but considerably more practical. You are looking at an industrial laser, which emits an extraordinary light, unknown in nature. It can project a spot on the moon. Or at closer range, cut through solid metal. I will show you."*

Goldfinger snaps his fingers, and a henchman turns on a machine. You can see a metal rod moving over Bond and then a red light beam emanating from its end. It hits the plate at the edge about one meter below Bond's "wedding tackle" and melts the gold there. Slowly the red laser beam moves upwards and threatens to divide the top agent.

Goldfinger: *"This is gold, Mr. Bond. [...]"*
Bond: *"I think you've made your point. Thank you for the demonstration."*
Goldfinger: *"Choose your next witticism carefully, Mr. Bond. It may be your last. The purpose of our two encounters is now very clear to me. I do not intend to be distracted by another. Good night, Mr. Bond."*
In the meantime, the laser beam has come quite close to the masculinity of the secret agent, and Bond shows slight signs of panic.
Bond: *"Do you expect me to talk?"*
Goldfinger: *"No, Mr. Bond. I expect you to die!"*

We now want to analyze this scene and especially the dialogue between Bond and Goldfinger in detail. So Goldfinger has a new toy, a laser beam, as he says. Today one would simply say a laser. What is so special about the laser light that this "toy" emits? To this end, we must first consider how light is generated. Physics knows only one mechanism here. When an electron absorbs energy in an atom, it changes to a higher energy state, as the physicist says. If this excited electron now releases its energy again, i.e., if it falls back into a deeper energy state, then this happens by emission of a light particle, the so-called photon. This is how light is created. So whenever you see light, electrons have to pass from higher energy states to lower energy states and emit photons. This process takes place chaotically in light sources such as our Sun, and also in light bulbs and candles. The light is therefore emitted in a completely uncoordinated way in all directions.

There is now a fundamental difference between the laser and this chaotic process. Atoms are excited in a laser medium and thus transformed from lower energy states into energetically higher ones, i.e., excited states. They are thus said to be "pumped". For certain special atoms, the pump energy remains stored there for a longer time, so that a population inversion can be built up, i.e., more electrons are in higher than in lower energy states. At this point the process of stimulated emission is triggered. Stimulation of an atom by a photon with precisely the energy to be emitted causes the excited atom to return to its ground state, emitting a photon with the same energy and phase as the stimulating photon. The two photons also move in the same direction. Due to this doubling of the stimulating photon, the laser medium acts like a light amplifier. The second photon then stimulates other excited atoms to radiate, and a chain reaction occurs. This arrangement is located in a resonator whose dimensions are matched to the desired wavelength. Thus a photon has many possibilities to stimulate other atoms when passing through the laser medium several times. The resonator is simply formed by two mirrors at the ends of the setup. The direction of the generated laser light beam is finally determined by these mirrors. One of the two mirrors lets some of the light through, and it can hit the gold plate, for example. Ultimately, this means that many electrons release their energy simultaneously in the form of light. The name "laser" is actually an abbreviation. It stands for *Light Amplification by Stimulated Emission of Radiation*. Goldfinger is actually right when he says that this light does not exist in nature. There is no known natural process that would produce the population inversion described above—why should it? Why should electrons voluntarily assume higher energy states and cause population inversion? Natural laser radiation has therefore not yet been discovered either on Earth or in the entire universe. It is also true that one can create a light spot on the Moon with a laser. This has often been done to determine the distance of our companion with high accuracy, because laser radiation forms a very narrow beam, in contrast to normal light, due to the process used to produce it.[32] So Goldfinger's remarks about his new "toy" are all true.

But can the laser really cut through the gold plate as easily as can be seen in the film? Must Bond and especially his Bond girls really fear for his "family

---

[32] Laser light is not only very intense, but also has many other properties, such as its coherence, its perfect polarization, and its extremely narrow beam, which distinguishes it from light from other sources.

jewels"? To decide this, we have to consider the power, i.e., the amount of energy per unit time that is transmitted by the laser light. Probably the best known example of such an output is the light bulb. Here the power, measured in watts, indicates how much electrical energy, measured in joules, is converted into light and heat per second. Commercially available light bulbs usually have an output of between 25 and 100 watts, but only about 5% of this is converted into light.[33] In order to estimate the performance of Goldfinger's "toy", we must determine how long the laser takes to melt or cut the gold plate. The energy required to melt a material such as gold is essentially calculated from the mass, the temperature difference between the ambient temperature and the melting temperature, and the heat capacity of the material. The heat capacity indicates how well a body can absorb and store energy in the form of heat. In addition, the heat of fusion must also be taken into account in the energy balance. This is the energy that must be applied to a body for the transition from the solid to the liquid state. In a solid body, the atoms are in a regular arrangement, the crystal lattice. The heat of fusion is needed to break up this lattice. The cut caused by the laser shows that material is partially melted and partially even vaporized. The cutting edges are also not very clean. The required energy is therefore somewhere between the amount needed to melt and the amount needed to evaporate the gold. To evaporate metal, in addition to the energy required to melt it, energy must be applied for further heating up to the boiling point and for the transition from the liquid to the gaseous state.

If all these processes are taken into account, a calculation shows that Goldfinger's laser must have a power between 5 and 50 kilowatts, depending on whether the cut is caused by a melting or an evaporation process. The large discrepancy between these two values is due to the high energy required to evaporate metals. For comparison, a typical small car has an engine power of approx. 40 to 50 kilowatts. It must also be taken into account that a laser, like a light bulb, does not convert all the electrical power supplied to it into light power. A realistic efficiency of 10% was always assumed for the calculations concerning the film scenes. This means that when the laser is fed with 1000 watts of electrical power, it emits only 100 watts of light. The rest is usually lost as heat.

From a technical point of view, it would be no problem nowadays to achieve the calculated power of the laser appearing in the film Goldfinger. In fact, industrial lasers with such powers and even higher are used to cut and weld

---

[33] This poor efficiency of 5% ultimately led to a ban on the production of such light bulbs in the EU in 2009.

metals. Unlike the laser used by Goldfinger, real cutting and welding lasers use laser light beyond the visible light in the so-called infrared range.[34] The reason for this is quite simple. Metal surfaces reflect visible light, which is why metals have that typical shine. However, the light of the laser should not be reflected during the cutting process, but absorbed so that the metal is strongly heated at the corresponding point. Infrared lasers are used because metal surfaces are poor reflectors of infrared light. Goldfinger may have avoided this problem by some special treatment of the gold surface. Or perhaps the high power of the laser immediately changed the surface to such an extent that the light was preferentially absorbed. In any case, the scoundrel will surely have used some trick…

But another question remains. Why is the laser beam visible as a bright red beam in the film? With a laser pointer only the point on the projection surface is visible and not the whole beam. Only when there is smoke in a room, for example, can the beam of the laser pointer be detected because it is reflected by the smoke particles. For Goldfinger's laser toy, on the other hand, a different explanation is offered. In his gold foundry, of course, there are no smoke particles, but there is so much gold dust flying around the room that the laser beam is very clearly visible.

It is also interesting to note that Theodore Maiman's first laser[35] was put into operation on May 16th 1960. For a long time people didn't really know what to do with it, and only insiders knew about this strange invention.[36] This changed abruptly when the laser was used in the movie *Goldfinger*. Then the world suddenly discovered what a laser is and what the special properties of laser light are. The film undoubtedly contributed greatly to the spread of this invention. It should also be noted that Ian Fleming wrote his book *Goldfinger* in 1959—one year before the invention of the laser. In the book there is also this scene with 007 tied on the gold plate. However, he is not threatened with a laser there. A circular saw is called in to do the job. This is also quite nice, but of course no comparison at all to the very threatening and modern laser.

As you can see later in the film, Goldfinger uses his laser to melt through Fort Knox's huge armored doors. The laser is brought in a small trolley. Our calculations show that this could actually be the case. The power of a mid-range car would be sufficient to operate a laser with which metal could be melted. However, the chronology is problematic here once again. It was not

---

[34] Like light, infrared radiation is electromagnetic radiation, but with a longer wavelength, just beyond the wavelength of red light.

[35] As early as 1954, Charles H. Townes realized the first maser, which worked in the microwave range according to the same principle as a laser.

[36] It was even mocked with the statement *"The laser is a solution in search of a problem"*.

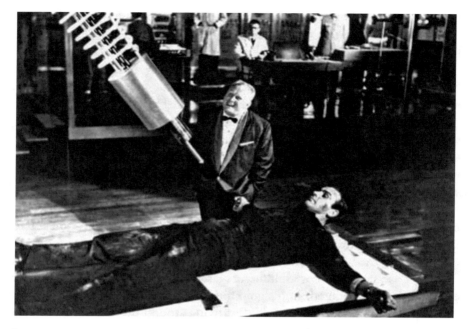

Figure 3.8   James Bond (Sean Connery) lies bound on a gold plate and is threatened by Goldfinger (Gert Fröbe) with a laser beam

until 1990 that lasers became so powerful that they could be used for welding and thus also for melting metal plates. Although lasers had been around for 4 years in 1964, they still didn't have enough power to do what was shown in the film. So there had to be tricks involved here. Sean Connery lay on a brass plate, as can be seen in Figure 3.8. This plate consisted of two halves that were soldered together in the middle. At the same time someone with a welding torch was lying under the plate and slowly moved it up along the soldered seam towards Connery's crotch. He couldn't see anything and just pushed the burner slowly on call.[37] Apparently, everything went well and Sean Connery was completely unharmed.

In the film, the laser beam finally comes dangerously close to Bond's "private parts". Desperately 007 mentions the top secret operation *"Grand Slam"*. The slightly confused Goldfinger then decides that James Bond could be more useful to him alive and switches the laser off at the last moment. Bond is stunned once more and taken aboard Goldfinger's plane.

---

[37] This was explained in the documentary *"Best ever Bond"* by Goldfinger director Guy Hamilton.

Details for Know-It-Alls

The definition of power as energy delivered per unit time allows us to find the desired laser power. The total energy E consists of the energy E_1 required for a temperature increase ΔT and the energies required for melting E_2 and evaporating E_3.

For E_1, we have $E_1 = c \cdot m \cdot \Delta T$ where m is the mass heated with a temperature rise of ΔT and c is the specific heat capacity, which is a constant for each material.

For E_2, we have $E_2 = m \cdot Q_{melt}$ with the heat of fusion Q_{melt}, which indicates how much energy is needed to melt one kilogram of a material. Q_{melt} is also a constant for every material.

A similar relationship applies to E_3, namely, $E_3 = m \cdot Q_{eva}$ with the evaporation heat Q_{eva}, which indicates how much energy is needed to evaporate one kilogram of a material.

Putting all this together, we obtain the total energy E needed to melt or evaporate the mass m of a material:

$$E = E_1 + E_2 + E_3 = c \cdot m \cdot \Delta T + m \cdot Q_{melt} + m \cdot Q_{eva}$$

The required laser power P is then simply $P = E/t$, where t is the time during which the laser beam acts on the material. This time is estimated to be quite small, at about one second, since the laser moves continuously. The values for the required laser power are then determined from estimates of the amount of molten or evaporated metal $m \approx 25$ g, the differences ΔT between the known melting and evaporation temperatures and the ambient temperature of about $20\,°C$, the melting heat $Q_{melt} = 63\ kJ/kg$ and the evaporation heat $Q_{eva} = 1650\ kJ/kg$ of gold, and the specific heat capacity of $c = 126\ J/(kg\cdot K)$ of the noble metal.

A Gun, a Plane, and Pussy Galore

After being stunned, James Bond wakes up in Goldfinger's plane—7000 meters over Newfoundland and guarded by pilot Pussy Galore. He's to be flown to Baltimore. Miss Galore announces the landing in 55 minutes. The smart secret agent then prepares himself in the bathroom. When he leaves the room with new clothes, Pussy Galore faces him with her revolver out. This dialogue follows:

Pussy Galore: *"We'll be landing in 20 minutes. Do you want to play it easy or the hard way? And this isn't a tranquilizer."*

James Bond: *"Pussy, you know a lot more about planes than guns. That's a Smith & Wesson .45. If you fire this close, the bullet will pass through me and the fuselage like a blowtorch through butter. The cabin will depressurize and we'll be sucked into outer space together. If that's how you want to arrive, you're welcome. As for me, I prefer the easier way."*

The question now is: Was James Bond just trying to unsettle Pussy Galore with his comments to keep her from threatening him any further, or is the scenario he describes actually realistic? (Figure 3.9).

First, let's take a closer look at the gun. According to 007 this is a Smith & Wesson .45. The number ".45" indicates the caliber of the weapon. This refers to the diameter of the bullet in the American inch unit. 0.45 inches is about 11.5 millimeters. Smith & Wesson (S & W) is the world's largest manufacturer of small arms. The film Goldfinger was shot in 1964. At this time there

Figure 3.9 A picture from the set: Pussy Galore (Honor Blackman), James Bond (Sean Connery), and the Smith & Wesson .45

were two S & W revolvers with caliber 45 in series production: on the one hand, the model S & W M 1917 and on the other, the model S & W 25-2. The latter has a barrel length of 152 millimeters and accelerates the projectile to an exit speed of 270 meters per second. This corresponds to almost 1000 kilometers per hour and is only about 20 percent below the speed of sound. The S & W M 1917 achieves "only" 213 meters per second, which is why the S & W 25-2 is better suited to achieve great penetration power. When Pussy Galore's weapon is compared to these two models, we find that she must have been in possession of an S & W 25-2.

It is interesting that a cartridge of the type .45 ACP (Automatic Colt Pistol) was usually used in the S & W 25-2. This was invented in 1905 by John Moses Browning and used by the US military for a long time. This is important for the film scene, because the ACP cartridge has a relatively large diameter in relation to the rather low kinetic energy. This is intentional because it normally gets stuck in human targets and causes serious internal injuries without endangering other people, such as soldiers from one's own camp or hostages to be liberated. However, we do see Pussy Galore pointing her gun at Bond's belly. If she fired her revolver, the bullet would only hit muscles and innards that offer less resistance than Bond's ribs.

Did James Bond exaggerate the effects of the shot just to save himself? First, we need to find out whether the bullet would indeed go through him and the plane wall, like a *"blowtorch through butter"*. If, for example, a pistol is used to fire at a block of wood, a firing channel is created. When the projectile hits the target, it must displace the wood and accelerate it outwards.[38] As early as the 17th century, the famous British physicist Sir Isaac Newton investigated how deeply a projectile would penetrate into a given material, i.e., how long this firing channel is. He came to the conclusion that at high projectile velocities[39] the exact value of the velocity does not matter at all. Newton found that the length of the firing channel is equal to the length of the projectile times the density ratio between the material of the projectile and that of the target.[40] Pussy Galore's bullet can easily make it over 100 meters without an obstacle, so the distance from the pistol to James Bond is negligible. The bullet would reach him at full speed. James Bond is well trained, and therefore about 25 centimeters thick, measured from belly to back. The pistol is aimed at the

---

[38] In the case of bullet holes in a metal plate, it is clearly visible that the material is pressed outwards in the direction of the bullet.

[39] This means that the velocity must be at least high enough for the projectile to penetrate completely into the material hit.

[40] A 1 centimeter long projectile that penetrates wood would thus create a 10 centimeter long firing channel, because the density ratio of the projectile material to wood is about 10 to 1.

abdomen, which contains no bones and has a significantly lower density than the chest with the ribs. A bullet of the .45 ACP cartridge type weighs up to 16.85 grams, corresponding to a lead cylinder with a length of 1.44 centimeters. The calculation then shows that the ball would penetrate 16 centimeters deep into the abdominal tissue and thus get stuck. This was also the intention when the weapon was made, because it was designed as a melee weapon, as already described.

The aircraft wall is made of aluminum and is about three to four centimeters thick. The density of the wall is slightly more than twice that of James Bond's belly. If the latter were not in front of the wall, the maximum penetration depth of the projectile would be almost six centimeters. So it would just smash through the wall. However, it would definitely be impossible for the bullet to go *"like a blowtorch through butter"*, traversing both Bond's body and the wall of the plane. Bond's testimony is thus exposed as a bluff to save his neck, or rather his stomach.

But suppose Pussy Galore actually missed Bond and shot a hole in the outside wall of the plane. Would that really be dangerous? When she targets him, James Bond says to her: *"The cabin will depressurize and we'll be sucked into outer space together."*

In fact, the air pressure usually maintained in an aircraft is similar to the air pressure at an altitude of 2000 meters. Since the airplane is flying far higher, there is thus an overpressure in the interior. So far, so good. Of course, a small hole in the outer wall would cause the air to escape from the inside. But it is also clear that the cabin cannot be hermetically sealed.[41] Every modern aircraft has different valves that are practically nothing more than controlled leaks. Fresh air is constantly supplied via the air conditioning system. If a leak occurs, for example because someone like Pussy Galore fires a shot at the outside wall of the plane, the valves will simply close a little further. The effects of a handgun are therefore far from sufficient to cause a plane to crash. The greatest danger from a leak is rather that it can lead to an oxygen deficiency. If a pressure loss occurs for any reason, the oxygen masks known to all are used. If a passenger in the aircraft puts these on within the next 15 seconds, there will be no damage to health. If the oxygen supply remains at too low a level for any longer period, a person will quickly go into a state in which they feel particularly well and think that they will continue to be able act in a controlled manner without the oxygen supply. This is a fatal error, because shortly afterwards, they will first lose consciousness and then death will occur.

---

[41] Otherwise, passengers would spend the whole flight in a cabin contaminated by human body odours. This would certainly be unpleasant during a long intercontinental flight.

And why does Bond say in the German version of this movie that they will be sucked *"out into the universe"*?[42] That's a little surprising, because space, the beginning of the rest of the universe, starts at a height of approximately 80 kilometers, and the plane is only seven to ten kilometers above the Earth's surface. Once again, this is one of those embarrassing translation errors which was made when the movie was dubbed into German. In the English original James Bond does not speak of *"into the space of the universe"*, but says *"into outer space"*. It's clear to everyone that, in this scene, *"into outer space"* means only *"outwards"*.

So we find that, in this scene, 007 was exaggerating about the effects that Pussy Galore's shot might have in the airplane. Firing a weapon would not be particularly dangerous for those on the aircraft. By far the greatest danger posed by an armed person in an aircraft is clearly that you could be hit by a bullet!

Details for Know-It-Alls

The penetration of a projectile into a target is found by a simple application of energy and momentum considerations in physics. If the projectile gets stuck, then the energy transferred to the displaced material is at most as large as the energy of the projectile. The energy of the bullet and the displaced material in the target depends only on their volume, velocity, and density. The speed enters into the formulas as a square. However, it is the same for the bullet and the material and therefore does not have to be taken into account. Therefore, the energies can be equated and the resulting equation solved for the penetration depth L:

$$L = L_0 \cdot \rho / \rho_{mat}$$

The density ρ of the bullet is easy to determine. It is made of lead, which has a density of $\rho = 11.3$ g/cm^3. It is also known that the ACP bullet commonly used in the Smith & Wesson .45 has a mass of up to 16.85 g. This results in a length of 1.44 cm for the given caliber of the weapon, if we assume a cylindrical shape for the projectile. The density of James Bond's belly is $\rho_{mat} = 1$ g/cm^3. This corresponds to the density of water. Since the human body consists of 80% water and the abdomen is mainly fat and connective

---

[42] In the German version "outer space" was mistranslated by "Weltraum" which means the space of the universe.

tissue, this estimate is certainly quite accurate. The density of the special aluminium used in aircraft is relatively low for a metal, with $\rho_{mat} = 2.7$ g/cm^3. In addition to aluminium, the wall also consists of cavities and insulating materials, which have a lower density, so the bullet would penetrate even further into them if it could go through 6 cm of aluminium. These values were used to calculate the penetration depths in the text. It should be noted, however, that the calculation only applies to high velocities at which the projectile penetrates completely, because otherwise the so-called cohesion energy cannot be neglected in relation to the kinetic energy.[43] If, for example, the bullet is only thrown against the plane wall, it will certainly not be going fast enough to penetrate, because the necessary cohesion energy is not available as kinetic energy. Since James Bond is close enough to the wall of the plane, it can be assumed that the bullet will reach the wall at a sufficient speed and at least penetrate the wall.

Operation "Grand Slam"

James Bond is taken to Goldfinger's ranch, where the villain's plan initially works out rather well. 007 is observed by the secret service, whose suspicions are not then aroused regarding the fact that he might be in trouble. That was exactly why Goldfinger spared the top agent from his laser and shipped him to Baltimore. There Goldfinger has just gathered his henchmen around him, including those who have lent him money to carry out his operation *"Grand Slam"*.[44] James Bond is held prisoner in the basement of the building and is able in part to overhear the plan. Goldfinger is obviously planning to break into some kind of bank. He explains an aerial photograph of a certain region to his fellows:

Goldfinger: *"This is my bank. The gold depository at Fort Knox, gentlemen. In its vaults are $15 billion. The entire gold supply of the United States."*
The gangsters are very skeptical, and Goldfinger pushes a switch so that the floor of the room changes and a model of the Fort Knox area appears.

---

[43] The cohesion energy is the energy required to remove a particle from a material.
[44] In sports, *"Grand Slam"* means that someone "wins everything". A Grand Slam in tennis means winning the four most important tournaments in a year. A Grand Slam in baseball is a home run when all four bases are occupied. In ski jumping, the Grand Slam means winning all four jumps in the Four Hills Tournament, and in eventing (horse trials), a Grand Slam means winning the three most important tournaments in a given year.

First gangster: *Cut the commercial!—Get to the point! It's pointless. The depository's impregnable.*"
Second gangster: *"The joint is bombproof, electrified …"*
Goldfinger: *"Bear with me, please! Fort Knox is a bank. Like any other. Larger, better protected perhaps, but nonetheless a bank! It can be—I think the expression is—blown. My plan is foolproof, gentlemen! I call it Operation Grand Slam."*

Fort Knox was built in 1862 and named after the first US Secretary of War Henry Knox. It is a base of the US Army and is located in the state of Kentucky. The Ministry of Finance established the Bullion Depository adjacent to Fort Knox in 1936 as a repository for part of its gold reserves. With 4580 tons of gold, Fort Knox is home to one of the largest gold reserves in the world.[45] Its value today is about 180 billion dollars. That's 12 times more than the 15 billion that Goldfinger mentions. However, that value must of course be related to the year 1964. In the 1960s, the price of gold was linked to the US dollar, which was regulated by the Bretton Woods system.[46] In 1964, the year the film was released, one troy ounce of gold cost 35 US dollars.[47] At that time, 10,500 tons of gold were stored there, considerably more than today. It had a total value of $12 billion which is close to the $15 billion mentioned by Goldfinger. Goldfinger is still a little off, but it's the right order of magnitude.

In the German translation of this film, however, Goldfinger mentioned that the gold had a value of 15 trillion US dollars. Could he really have miscalculated its value by more than a factor of 1000? The answer, of course, is no! In the original English version, Goldfinger correctly speaks of "15 billion", as mentioned before. This was wrongly translated to "15 trillion". So this is once again an extremely embarrassing translation error, because pupils already learn at the German high school that the English word "billion" is not the same as the German word "Billion", which actually means "trillion".

Operation *"Grand Slam"* is one of the best but also least detailed actions of a James Bond villain. The gold-smuggler Auric Goldfinger obviously plans to break into Fort Knox. That's for sure. But what exactly does he want? Does he want to steal the gold? In the novel Goldfinger by Ian Fleming this is indeed the case.[48] But in the film the authors actually came up with something far more perfidious, as we shall see later. 007 has already got some first insights

---

[45] There is even more gold in the Federal Reserve Bank of New York. However, gold is also stored here by foreign banks and central banks.
[46] Gold has been freely traded since the collapse of the Bretton Woods system in 1973.
[47] One troy ounce (or ounce) corresponds to 31.1 grams.
[48] There the gold is to be transported away with a freight train. If each wagon can carry 200 tons of payload, one would need a train with 50 wagons!

into this plan, because he was able to eavesdrop on Goldfinger from his cellar prison when the villain was explaining Operation *"Grand Slam"* to his fellow criminals.[49] However, the top agent does not yet fully understand Goldfinger's machinations. He is taken to the veranda in front of the main house. Goldfinger is sitting in the shade with a drink, and at the beginning of the conversation James Bond also has a drink brought in. Bond faces Goldfinger and waits for his drink. Some interesting facts and figures are mentioned, which explain Operation *"Grand Slam"* in more detail. The dialogue begins like this:

Bond: *"You disappoint me, Goldfinger. You know Operation Grand Slam simply won't work. And incidentally Delta-9 nerve gas is fatal."*
Goldfinger: *"You are unusually well informed, Mr. Bond."*
Bond: *"You'll kill 60000 people uselessly."*
Goldfinger: *"Ha! American motorists kill that many every two years."*

We stop the dialogue here again for a fact check. First of all it should be made clear that there is no nerve gas with the name "Delta-9".[50] Secondly, let us ask whether the number of deaths on the roads in 1964 could actually have been that high, even though there was only a fraction of the traffic we see today. But also, how does Bond come to the number of 60,000 people? The second question is relatively easy to answer. This figure refers to the personnel of the military base at Fort Knox, if family members living there are included. If Goldfinger wants to break into Fort Knox and steal the gold supply of the United States, he must of course eliminate all these people. The number of road fatalities is also interesting. In fact, in the 1950s, this figure was around 35,000 deaths per year. So that's about 70,000 in two years. However, in the 1960s this figure rose steadily as traffic density increased. By 1970, 55,000 people died on American roads every year. It was not until the mid-seventies that the number of road deaths began to fall with the introduction of the seat belt. In 2016, however, this figure is still around 40,000, which corresponds to the current level.[51] Thus the number 60,000, which Goldfinger mentions for two years, corresponds perfectly with the reality at that time and even today, although it is slightly on the low side.

---

[49] It's a little surprising, though, that Goldfinger kills all his gangster colleagues shortly after he's explained his plan in detail.

[50] In chemistry, the substance tetrahydrocannabinol is referred to more precisely as delta-9 THC. However, this is cannabis, i.e., a solid substance and not a gas. Cannabis may be unhealthy, but it is not so toxic that inhaling small amounts will lead to immediate death.

[51] In Germany there were 3200 deaths in the same year. What is striking is that this figure per inhabitant is only about half as high as in the US, despite its rigid speed limit. In fact, about 50 percent of American road deaths in 2016 were without a fixed seat belt.

Sitting in an armchair opposite Goldfinger, Bond then continues: *"Yes, well … I've worked out a few statistics of my own. $15 billion in gold bullion weighs 10,500 tons. 60 men would take 12 days to load it onto 200 trucks. At the most, you'll have two hours before the army, navy, air force, and marines move in and make you put it back."*

Let us check these figures immediately. There is again a horrible mistake in the German version of the film. Bond speaks here about *"$15 million in gold bullion"*. Why is Bond talking about $15 million? Has the gold price fallen so suddenly, because shortly before Goldfinger was still talking about $15 trillion? Of course not! In the original English version James Bond says correctly "15 billion", as mentioned above. This number can once again be put down to the extremely sloppy dubbing of this blockbuster.

Since 200 trucks are available for the gold, each will have to be loaded with 53 tons. This is certainly possible, although 53 tons per truck is quite a lot.[52] The large Liebherr T282 dump truck, for example, has a maximum payload of slightly more than 350 tons, i.e., far more than 53 tons. A simple division shows that each of the 60 men will have to move about 14.6 tons in one day. This corresponds to 49 gold ingots of 12.44 kilograms every hour. But since there is even an elevator in Fort Knox and one can use aids such as small wheelbarrows for transportation, this will certainly be easy enough for Goldfinger's rogue troops to carry out. James Bond's number games are therefore realistic. In particular, however, he is absolutely right to say that doing this in two hours would not be feasible. Even a super villain can't have 5000 tons of gold removed every hour!

The conversation between Goldfinger and James Bond continues like this:

Goldfinger: *"Who mentioned anything about removing it?"*
He takes a break and looks into his glass, while Bond drinks a sip.
Goldfinger: *"Is the julep tart enough for you?"*
Bond: *"You plan to break into the world's largest bank but not to steal anything. Why?"*
Goldfinger: *"Go on, Mr. Bond."*
Bond: *"Mr. Ling, the Red Chinese agent at the factory? He's a specialist in nuclear fission … (thoughtfully) … But of course! His government's given you a bomb."*
Goldfinger: *"I prefer to call it an atomic device. It's small but particularly dirty."*
Bond: *"Cobalt and iodine?"*
Goldfinger: *"Precisely."*

---

[52] However, it should be noted that James Bond certainly does not calculate in metric tons. In the United Kingdom, the ton weighs 1016 kilograms.

Bond: *"If you explode it in Fort Knox, the entire gold supply of the United States will be radioactive for…* (thinking) *… 57 years."*
Goldfinger: *"58 to be exact."*
Bond: *"I apologize, Goldfinger. It's an inspired deal. They get what they want—economic chaos in the West. And the value of your gold increases many times."*
Goldfinger: *"I conservatively estimate ten times."*

So now we are talking about the truly ingenious core of the "Grand Slam" enterprise. Goldfinger wants to break into a bank, but steal nothing—who would have thought! Goldfinger wants to contaminate Fort Knox radioactively for 58 years with a "nuclear device". The gold stored there would therefore not be accessible during this time. In addition, this atomic device, which seems to be nothing more than an atomic bomb, is supposed to contain cobalt and iodine and be particularly "dirty". How does it all fit together?

Can gold be activated, i.e., made radioactive, for 58 years? An activation of gold is possible in principle, for example, by bombardment with neutrons. However, such a procedure has one serious disadvantage: the gold irreversibly decomposes into other chemical elements. But of course this must not happen, because Goldfinger loves gold. He would never destroy it. So we can be sure that's out of the question. We are not talking about a nuclear bomb at all, but about a "dirty bomb". This means that it is not a nuclear bomb in the classical sense, based on nuclear fission, which releases many neutrons, but a conventional explosive bomb. This explosive then distributes stored radioactive material in the environment, thus contaminating a large area. In our case, this obviously means that the dirty bomb is supposed to distribute radioactive cobalt and iodine in the room where Fort Knox gold is stored. The explosive power of such a bomb would also leave most of the gold completely undamaged. If an atomic bomb were detonated, on the other hand, the gold would simply evaporate. This is another reason why a classical atomic bomb cannot be used as the atomic device. So the question now is whether such sprinkling by a "dirty bomb" with radioactive cobalt and iodine would really be enough to make the gold inaccessible for such a long time?

First we need to find out what's inside the bomb. A glance at an isotope table[53] (see Table 3.1)[54] shows that most isotopes of cobalt and iodine are either very short-lived, i.e., have half-lives in the range of a few days at most,

---

[53] Each element of the periodic table has a well defined number of protons; cobalt, for example, has 27. This is the atomic number of an element. The number of neutrons, on the other hand, is not fixed, so there are different possible nuclear compositions, called isotopes. Isotopes of cobalt are for example ^{59}Co, ^{60}Co, ^{61}Co, etc.

[54] Isotope table adapted from http://ie.lbl.gov/toi .

Table 3.1 Extract from an isotope table for cobalt and iodine

Cobalt			Iodine		
Isotope	Half-life	Decay type	Isotope	Half-life	Decay type
^{55}Co	17.53 h	β^+	^{123}I	13.27 h	β^+
^{56}Co	77.27 d	β^+	^{124}I	4.18 d	β^+
^{58}Co	70.86 d	β^+	^{126}I	13.11 d	β^+
^{59}Co	stable		^{127}I	stable	
^{60}Co	5.27 a	β^-	^{128}I	24.99 min	β^+, β^-
^{61}Co	1.65 h	β^-	^{129}I	1.57×10^7 a	β^-
^{62}Co	1.50 min	β^-	^{130}I	12.36 h	β^-
^{63}Co	27.4 s	β^-	^{131}I	8.02 d	β^-
^{64}Co	0.30 s	β^-	^{132}I	2.30 h	β^-

The half-life indicates the time after which exactly half of the original amount of a radioactive substance is still present. In addition to the usual hours (h), minutes (min), and seconds (s), other time units needed here are days (d) and years (a). The type of decay refers to the scheme according to which the radioactive substances decay. In most cases, these are beta emitters that emit electrons or positrons as radioactive radiation

or very long-lived, with half-lives of several million years, an unimaginably long time. The only matching element is cobalt 60, written ^{60}Co, with a half-life of 5.27 years.[55] This means that after 5.27 years only half of the original amount of the substance is still present. The rest will be then have decayed to another stable element. There is no suitable isotope of iodine. Here all half-lives are either much too short or much too long. Iodine is therefore probably present in the atomic device for other reasons.

The atomic device thus contains large amounts of radioactive cobalt 60 isotope with a half-life of 5.27 years. After 58 years, only one out of every 2000 atomic nuclei of the original amount of material would still be present. In principle this sounds reasonable, but would the material really be safe again after 58 years? To answer that, we must first determine the initial amount of radioactive cobalt.

It can be assumed that the explosives are stored in the lower part of the bomb. At the end of the film, we actually get to see the atomic device. We see two spheres from the back, and from other perspectives we also see a third one. These are apparently completely filled with cobalt 60, except for a protective layer of lead two centimeters thick. A protective layer is absolutely necessary because the radioactive radiation is already present before the bomb is detonated. The radiation is thus at least somewhat shielded. However, it would still not be particularly healthy to stand for too long next to such a device.

---

[55] Cobalt 60 refers to the cobalt isotope, which contains 27 protons and 33 neutrons in the atomic nucleus.

Table 3.2 Absorbed radiation dose and its direct effect on humans

Radiation dose	Effects
0.002–0.004 grays	Annual radiation dose absorbed by humans as natural background radiation
0.5 grays	Radiation hangover: nausea, vomiting, headaches and dizziness
1 grays	Distinct symptoms: Fever, diarrhoea and bleeding, ulcers in the mouth and throat, hair loss, haematosis, infections
2 grays	50% of those affected die within 30 days
7 grays	Fatal within a few days
50 grays	Instantaneous death

Long-term changes, such as an increased risk of cancer, are not taken into account. The times given are based on the assumption that no medical care is provided

The amount of cobalt 60 originally present then results directly from the radius of the cobalt sphere. If the known body size of James Bond is related to the dimensions of the box, and these dimensions are related to the diameter of each cobalt ball, then this diameter can be determined quite accurately. This results in a value of 36 centimeters. The total volume of the three spheres is then calculated directly from this, and with the known density of cobalt, it follows that there are at least 460 kilograms of radioactive material present. We then proceed from the assumption that, after the detonation of the "atomic device", the radioactive material will be distributed more or less evenly over the floor of the room in which the gold is located. The size of this room can be estimated quite well from the film footage, giving a value of about 800 square meters.

Now we have to consider how heavily this room would be contaminated with 460 kilograms of cobalt 60. Therefore, we must now discuss radiation doses. Everyone is exposed daily to a very small natural dose, most of which comes from space as cosmic radiation. However, more than a thousand times this radiation would have to be absorbed for the organism to react with mild symptoms such as headaches and nausea. This happens in the range of 0.5 grays.[56] Corresponding effects for higher radiation doses are listed[57] in Table 3.2. The table shows that, for radiation doses less than two grays, it would still be possible to enter a radioactively contaminated room. The radiation in such a room would of course be extremely harmful to health. A dose of 50 grays, on the other hand, would immediately be fatal.

---

[56] A radiation dose of one gray is reached when one kilogram of matter absorbs a radiation energy of one joule. The unit known as the sievert is more commonly used. This also takes into account biological efficacy. For beta radiation, however, the units gray and sievert are identical.

[57] Source: http://lexikon.meyers.de/meyers/Strahlenschäden

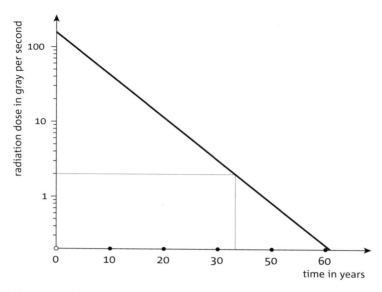

Figure 3.10 Decay of the radioactive radiation caused by 460 kilograms of cobalt 60 in the 800 square meter room with the gold bullions. The radiation dose is given as the number of grays to which one person is exposed per second. The life-threatening dose of two grays is indicated by the grey horizontal line. The grey vertical line shows that, after 33 years, this dose is still reached within one second in the contaminated room. The radiation dose is represented on a logarithmic scale as a function of time

It is well known that during radioactive decay the radiation decreases continuously over time. As Figure 3.10 shows, 460 kilograms of cobalt 60 reach a dose of two grays per second after about 33 years.

Since any meaningful cleaning work would take at least a few seconds, one could only start decontaminating the gold much later. In 1964, there were no remote-controlled robots that could have done this job for humans. Only after more than 50 years would the radiation dose have dropped to below one gray per second, which would then make it possible to start cleaning up. But even then one could not stay in the room for long periods of time without endangering one's health. Therefore, the contamination period of 58 years mentioned by Goldfinger is quite realistic.[58] Fifty-eight years after the detonation of the atomic device, the gold in Fort Knox would actually be accessible again.

---

[58] However, we have not yet explained why Bond initially calculates 57 years for the contamination of the gold, while Goldfinger then corrects him to 58 years. Since there is no fixed limit from which it can be said that the room could be entered again, 57 years would also be quite possible. This must be a tactical finesse on the part of 007, who wants to lengthen the dialogue as much as possible in order to elicit even more details about Operation *Grand Slam* from Goldfinger!

So this is the plan for Operation Grand Slam. Goldfinger seems to have thought of everything. What is surprising, however, is that Goldfinger wants to commit such a colossal crime just to increase the value of his gold and his wealth tenfold. He could just as easily have speculated on the stock market ...

Details for Know-It-Alls

Radioactive decay can produce three fundamentally different types of radiation: alpha, beta, and gamma radiation. Only very heavy nuclei, such as uranium or polonium, disintegrate with the emission of alpha radiation. A nucleus splits off two protons and two neutrons as a single entity, the helium nucleus. This results in a new atomic nucleus with an atomic number that is two smaller than the atomic number of the original nucleus, and this releases a lot of energy. Since helium atoms are relatively large charged objects, they cannot penetrate deep into matter.[59] A few centimeters of air, a sheet of paper, or even the skin are enough to completely shield from alpha radiation. If the gold had been contaminated with an alpha emitter, this would have had hardly any effect.[60] It could simply be retrieved from the room at Fort Knox and cleaned afterwards.

Beta radiation consists of electrons or positrons.[61] Since electrons and positrons are much smaller than helium nuclei, for example, they are able to penetrate deeper into matter, such as the skin. At low intensities, beta radiation can still be effectively shielded with a relatively thin layer of lead. But with the high radiation intensity that would prevail in the room at Fort Knox after the detonation of the atomic device, this would only be of limited help. With 460 kg of radioactive cobalt 60, it would certainly be difficult to completely shield from such beta radiation. However, wearing a thick lead apron it would still be possible to decontaminate the gold.

The third type of radioactive radiation is gamma radiation. Gamma rays are very short-wave and therefore extremely high-energy X-rays, and it is difficult to shield from this even with thick lead plates. Cobalt 60, the material Goldfinger uses for his atomic device, is initially only a beta emitter, as can be seen from a glance at Table 3.1, but secondary gamma decays occur with a

---

[59] Big is meant relatively here. The atomic radius for a helium atom is hardly more than one angstrom, i.e., a tenth of a billionth of a meter.

[60] It should be noted that alpha radiation is by no means harmless. If alpha emitters enter the body with food, for example, then their damaging effect is about twenty times greater than that of beta or gamma radiation.

[61] Positrons are positively charged electrons. When positrons are emitted, we speak of beta-plus (β+) radiation, otherwise in the case of electrons, of beta-minus (β-) radiation (see Table 3.1).

very short half-life. Cobalt 60 must therefore be treated as an effective gamma emitter with a half-life of 5.27 years, and the contamination of the room is even more dangerous than previously assumed. Even thick lead aprons won't help here anymore.

Since radioactive nuclei usually decay into stable nuclei, the radioactivity decreases continuously over time. It is easy to see that the number of radioactive decays is proportional to the initial number of atoms present, i.e., to the amount of material present at the beginning. This relationship is ultimately the cause of the exponential law that describes radioactive decay:

$$N(t) = N_0 \cdot \left(\frac{1}{2}\right)^{t/T}$$

Where $N(t)$ is the number of atoms that have not yet decayed at time t, N_0 is the number of atoms that were present at the beginning, and T is the half-life, i.e., the time after which only half of the originally existing atoms are still there. The half-life is a characteristic constant for every material and cannot be influenced by anything else. Suppose there are 1000 atoms of cobalt 60. Cobalt 60 has a half-life of 5.27 years. After this time there will only be 500 atoms left, after $2 \cdot 5.27 = 10.54$ years, there will only be 250 cobalt atoms left, after $3 \cdot 5.27 = 15.81$ years, only 125, and so on. After 10 half-lives, only $\left(\frac{1}{2}\right)^{10} = 1/1024 \approx 1$ thousandth of the initial substance will still be present. Ten half-lives makes about 53 years in the case of cobalt 60, which is almost the same as the 58 years Goldfinger mentions. After this time, 460 grams of the original 460 kilograms of cobalt 60 will still be there. Distributed over the 800 square meters of the room where the gold is stored, this is indeed a considerable reduction.

The activity of a radioactive material is the number of decays per second. The activity decreases just as exponentially as the total number of atoms.[62] A single cobalt 60 decay releases an energy of $E_{decay} = 5 \cdot 10^{-13}$ joules. Although this energy is very small, it must still be multiplied by the activity A. Thus the total energy released per second by the decay can be found from $E = A \cdot E_{decay}$.

The initial number of atoms N_0 can be calculated from the mass of cobalt in the atomic device of $m = 460$ kg. Since one mole of cobalt corresponds to 60 g and contains $N_A = 6.022 \cdot 10^{23}$ particles,[63] we thus have $N_0 = (460/0.06) \cdot N_A = 4.6 \cdot 10^{27}$.

---

[62] For experts, this is obtained immediately by differentiating the formula for the number of atoms with respect to the time.

[63] The number N_A is the Avogadro constant. It indicates how many atoms there are in a mole of a substance. One mole of a substance is the amount that corresponds to the atomic mass in grams. For cobalt

With this number, we can calculate the activity and thus the total radiation energy released per second in the 800 m² room.

If we consider that a person has a surface area of about one square meter and James Bond weighs 76 kg, then the result must still be divided by 800 and by 76 to arrive at the final radiation exposure in units of grays per second. The result can be seen in Figure 3.10, where the expected exponential relationship is converted into a straight line by the logarithmic scale.[64]

How Does Poison Gas Spread?

So now we know Goldfinger's diabolical plan. He wants to use Delta-9 poison gas to eliminate the 60,000 people guarding Fort Knox and then radioactively contaminate the space in which the gold is located to such an extent that it will no longer be accessible for a long time.[65] The effectiveness of Delta-9 can be seen when Goldfinger locks up his gangster bosses, those who had supported him financially until then, in an airtight room. The gas flows out from a small pressure bottle in the ceiling, and about ten seconds later the men collapse at the other end of the large room, about eight to ten meters away.

Even faster seems to be the gas that Pussy Galore uses with her squadron to take out the military garrison of Fort Knox. Several times we see entire columns of soldiers collapse, literally in the same second that an airplane flies over them. The planes are located at a flight altitude of approx. 50 meters, as can be seen by comparison with the large buildings in the film. The air seems to be calm, at least there is no sign of strong wind. Could it be that gas spreads so quickly in calm air? If, for example, a strongly smelling cheese is unwrapped somewhere, we would expect it to take a while for the odour to spread throughout the room. Why is that?

First it is important to realize that even the apparently calm air in a closed room is anything but calm. The molecules that make up the air, mainly nitrogen and oxygen, move at great speed, and the higher the temperature, the faster they move. At a room temperature of 20 °C, the speed is several hundred meters per second. For comparison, a Formula 1 car achieves 360 kilometers per hour on a long straight, which is just 100 meters per second. If the

---

60, one mole is therefore an amount of 60 g.

[64] While an interval on a linear scale always indicates constant distances (1; 2; 3; …), an interval on a logarithmic scale increases with powers of ten (… 0.01; 0.1; 1; 10; 100; …).

[65] Once again, there is a crucial difference with the book *Goldfinger*. In the novel, the soldiers are to be killed by poisoning their tap water. The question of why poison gas was used in the film can only be speculated upon.

air molecules, and with them the molecules that cause the smell or transport the poisonous gas, move this quickly, they will be able to spread very rapidly, won't they? Then, in fact, for the 50 meters from the planes to the ground, they would need only a fraction of a second, so the soldiers would be killed almost instantly.[66]

However, something very important has not yet been taken into account here. The molecules have a very high speed, but they never get very far, because they always collide with other molecules after a very short distance and are subsequently deflected in a completely different direction. So they don't move in a straight line, but in a zigzag. It is easy to calculate the distance they are likely to cover unhindered. In one cubic millimeter of air, a tiny cube of one millimeter edge, 27 quadrillion molecules[67] crowd together; written out, that would be 27,000,000,000,000,000. Each individual molecule thus has only a very limited volume freely available to it for motion. This free space can be understood as a kind of tunnel between the other molecules, through which the given molecule has to pass. The length of the straight line along which the molecule can move freely can be calculated from the required cross-section of the tunnel and the available volume. It is about one ten-thousandth of a millimeter. Since the flight speed of a few hundred meters per second is known, it is possible to calculate how long the free flight of a molecule will last, or again, how often it will collide with one of the many other molecules every second. That's a few billion times, after all. Even the most stale air in a long closed room has a lot of dynamics to offer.

Let us go back to the question of how molecules (poison gas or cheese smell) spread in air. They collide again and again with other molecules and change their flight direction each time. In the course of time they therefore move away from their starting point, or perhaps return somewhat, doing this again and again. Exact predictions for the motion a given molecule cannot be made. However, it is clear that in the course of time the molecules will be found increasingly far away from their starting point. Such a motion is called diffusion, and the corresponding theory of molecular propagation has been known in physics for two centuries. We can show how such motion works with a small computer simulation.

Each of the three images in Figure 3.11 shows the motion of ten molecules that start in the center of the image and fly the same fixed distance straight ahead in each step, but change their flight direction randomly after each step.

---

[66] That would also mean that the gangster bosses Goldfinger killed with the gas earlier would have died even faster than shown in the film.

[67] That's a really big number: if you could sell the molecules at a price of one cent for every 1000, then the total price would be about as high as the annual German federal budget.

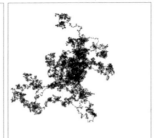

Figure 3.11 The zigzag course of ten gas molecules that start in the middle of the image and collide again and again with other molecules that are not visible. On the left you can see the course after 1000 collisions for each molecule, in the middle after 4000 collisions, and on the right after 16,000 collisions. The molecules spread through space in the process known as diffusion

In the first picture each molecule has made 1000 such steps, in the second 4000, and in the third 16,000. From the pictures we can see what is confirmed by a more exact calculation: in order to move twice as far away from the starting point, a molecule does not need twice the number of steps, but four times the number, and hence four times the time. For a tenfold distance, a hundred times more time is already needed. The propagation of a gas by diffusion therefore becomes slower and slower as it progresses.

How fast the diffusion takes place also depends on the type of gas, in particular on the size and mass of the molecules. As a concrete example, for benzene, a medium-sized and medium-heavy molecule, we can calculate how long it takes for the molecules to reach a certain distance from the source. An average distance of one meter is only reached after a good five hours; ten centimeters, after just three minutes. A benzene molecule, on the other hand, needs no fewer than 22 days to cover a distance of ten meters by diffusion!

Figure 3.12 shows the concentration of a gas propagating from a point in space after being suddenly released there. After a certain period of time, the concentration is distributed depending on the distance to the central point, as shown by the solid line. After four times the time, the dashed distribution of the gas results. The two dots mark the mean distance of the gas molecules from the starting point. After four times the time, this distance has doubled, exactly as the computer simulation has indicated.

Anyone who now objects that cheese unwrapped on the table does not only start to stink after five hours is of course right. There are two reasons for this. Firstly, even if the average distance of the molecules from their source is only one meter, there will be many molecules that are already much further away at the same time. And very strongly smelling or very poisonous substances

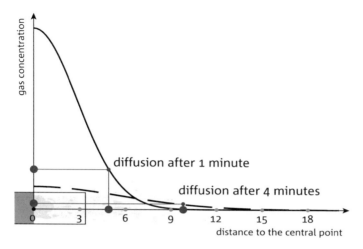

Figure 3.12 Dispersion of a gas by diffusion in space. The curves show the gas concentration as a function of the distance from the central point where the gas was released. The solid curve shows the situation a certain time after release of the gas, e.g., after one minute. The dashed line indicates the concentration distribution in space after four times the time. The two black dots indicate how far the gas molecules have moved on average from the starting point at these two times. For four times the time, the mean distance doubles

also unleash their effect in small concentrations. Secondly, the above figures apply only when the air is absolutely still. In a room in which there are people, which is heated or ventilated or illuminated by the Sun, there is always a slight flow of air which ensures that molecules can spread faster than in pure diffusion. A barely perceptible draught with a speed of only ten centimeters per second would, for example, ensure that the smell of cheese could get a meter away after just ten seconds.

For the propagation of gases over distances of one meter or more, air flow will certainly have more influence than pure diffusion. The scene from Goldfinger in which the gangsters are locked up and poisoned is therefore somewhat problematic. The speed at which the gentlemen collapse suggests a rather violent draught of at least one meter per second in the room, but this is not really compatible with the fact that all the doors and also the chimney are perfectly sealed to prevent the gas from escaping. Here the air in the room is obviously being vigorously moved around by a fan which is not visible.

Regarding the spread of poison gas by Pussy Galore's squadron, the scene is highly unrealistic. In order to bring the gas from the altitude of the aircraft to the ground so quickly, there would have to be a violent storm with winds of at least 50 meters per second, i.e., 180 kilometers per hour, but there are no signs of this at all. The flowers in the front gardens and the branches on the trees are only swaying rather gently. It can also be seen that the planes are flying very close to each other. If the gas were to spread so quickly, the pilots would inevitably end up poisoning each other.

So it should be borne in mind that, if people drop dead immediately at some distance, then there must always be quite a strong draught in the room. Pure diffusion, which would be the only mechanism for the propagation of molecules in perfectly calm conditions, is a slow process and could never lead to the instantaneous death of people in the vicinity.

However, the fact that the soldiers fall down immediately when the poison gas is sprayed has a completely different reason in the film. They only fake their death, because 007 has brought Pussy Galore over to his side using his irresistible qualities shortly before her mission and the deadly poison gas has been replaced with something else. The seemingly instantaneous death of the soldiers should have made Goldfinger suspicious, but otherwise Operation *Grand Slam* appears to be proceeding according to plan. Goldfinger's henchmen penetrate the Fort Knox gold chamber with a laser and place the atomic device there. They are then disturbed by the suddenly awakening troops and there is a fierce battle between the two parties. Goldfinger loses this fight, but manages to escape, and the nuclear device is defused in the end, by James Bond of course, only seven seconds before it is due to explode.[68] It could have made Auric Goldfinger fabulously rich—if it had actually detonated.

Details for Know-It-Alls

Heat is nothing else than the kinetic energy of molecules. The absolute temperature T of a gas measured in kelvin is related to the mean velocity v of the gas molecules by

$$\tfrac{1}{2} \cdot m \cdot v^2 = 3/2 \cdot k_\mathrm{B} \cdot T.$$

The left-hand side is the kinetic energy of a molecule of mass m. The Boltzmann constant $k_\mathrm{B} = 1.38 \cdot 10^{-23}$ J/K plays an important role in all many-particle phenomena. If the mass of a nitrogen molecule[69] N_2 is used, i.e., 28

---

[68] The atomic device has a three-digit time display which counts down the seconds until detonation. At the end this counter actually shows "007"!

[69] Nitrogen is the most abundant component of the air, accounting for 78% of the total. The 21% oxygen molecules are slightly heavier than the nitrogen molecules.

atomic mass units or $28 \cdot 1.67 \cdot 10^{-27}$ kg, then for a temperature of 293 kelvin (i.e., 20 °C), the result is an average speed of about 500 m/s or 1800 km/h.

Under normal conditions (atmospheric pressure, 0 °C temperature), one mole of any gas will occupy a volume of 22.4 liters, i.e., 0.0224 m³. One mole contains $N_A = 6.023 \cdot 10^{23}$ molecules. This is the Avogadro number again, and the results is $2.7 \cdot 10^{16}$ or 27 quadrillion molecules per mm³. The reciprocal of this number is just the volume available per molecule.

The volume per molecule is linked to the mean free path length L, i.e., the unhindered flight distance between two collisions that a molecule can expect in the air. Every molecule has a certain size. For the sake of simplicity, the molecules can be imagined as small spheres of radius r. In order for the molecule to avoid colliding with another molecule on its flight, there must be no other molecular center at a distance $2 \cdot r$ from the trajectory of the center of the molecule. In the case of a free flight of length L, this means that no other molecule may be present in a cylinder with length L and radius $2 \cdot r$. The volume of this cylinder is $4 \cdot \pi \cdot r^2 \cdot L$. If this volume is equated with the volume available per particle, then the mean free path length L *is* obtained. The result is one ten thousandth of a millimeter for nitrogen in air.

The diffusion of a substance is described by the so-called diffusion equation.[70] It has been known for over 200 years and can be solved for simple situations without a computer. The result is that the distribution of a substance diffusing from a very small source[71] into a very large space is described by a Gaussian bell curve[72] (see Figure 3.12 solid line). The width of this curve increases proportionally to the square root of the time that has passed since the sudden release of the substance (see Figure 3.12 dashed line). The height of the curve, i.e., the concentration of the substance at the starting point, decreases accordingly, since the material is distributed further and further into the surrounding space.

Out into Space …

At the end of the film, James Bond is on board a Lockheed Jetstar plane at 10,000 meters altitude on the way to the White House in Washington. The American President would like to thank him for saving the United State's gold

---

[70] The same equation also mathematically describes the process of heat conduction.

[71] Strictly speaking, we should speak of a pointlike source from which a substance spreads into an infinitely large space.

[72] The formula for the Gaussian bell curve and the corresponding graph used to be shown on the old German ten mark bills.

reserves from radioactive contamination. He is enjoying a nice relaxed flight and has just released the seat belt when Auric Goldfinger enters the passenger compartment of the aircraft holding a gold-plated Smith & Wesson 25-2 caliber .45 in his right hand and wearing the uniform of a general. The villain has already overwhelmed the other passengers and taken over the plane. 007 is surprised, and the following short conversation takes place (Figure 3.13):

Goldfinger: *"I'm glad to have you aboard, Mr Bond."*
Bond: *"Well, congratulations on your promotion, Goldfinger. Are you going to the White House, too?"*
Goldfinger: *"In two hours I shall be in Cuba. You have interfered with my plans for the last time, Mr. Bond."*
Bond: *"It's very dangerous to fire guns in planes. I even had to warn Pussy about it. By the way, where is she?"*
Goldfinger: *"I will deal with her later. She is where she ought to be—at the controls."*

Bond exploits a brief moment of carelessness on the part of his assailant and tries to take hold of the gun. There is a brief struggle with Goldfinger during which a shot is fired and the bullet hits a window a few meters away. The glass shatters and the plane immediately loses altitude. While the pilot

Figure 3.13 Goldfinger (Gert Fröbe) threatens James Bond (Sean Connery) with a revolver in an airplane

tries to stabilize the plane and James Bond clings to a luggage rack, Auric Goldfinger loses his grip and is sucked toward the broken window together with various objects, including a floor lamp and a sofa cushion. He gets stuck in the window for a moment, but then gets drawn out of the plane legs first. Could this really happen?

Knowing the proportions of 007's head compared to the airplane window, we can estimate its width and height. The window seems to be square and we find 40 centimeters for the width and height. The shot destroys the window and creates a suction toward the outside. This happens because every passenger plane has a pressurized cabin. At 10,000 meters altitude, the cruising altitude of most aircraft, the air is very thin. Indeed, the oxygen content and the air pressure are much lower than on the ground. Only after intensive training can mountaineers climb up to eight thousand meters without breathing apparatus. Even further up, survival would hardly be possible, especially for untrained passengers and flying personnel. Therefore, the air pressure in the parts of the aircraft where there are people is maintained at a level equivalent to the air pressure at an altitude of 2000 meters—a level that most people do not yet experience any problems with. At cruising altitude there is therefore a large pressure difference with the external environment of the aircraft. With a large leak in the outer wall, the air can then naturally flow from the inside to the outside, in such a way as to equalize the pressure.

Thermodynamic systems, for example, two rooms containing different gases and with a connection between them always try to adapt their conditions to each other. This is also the reason why there is wind and weather: different places in the atmosphere have different air pressures, and the air flows from the high pressure area to the low pressure area to compensate for this difference. This air flow is what we experience as wind. So there will be an air exchange in the film scene, too, when the window gets smashed, and this exchange will continue until the pressure inside and outside the plane is the same. Since the pressure outside is lower, the air flows outwards. In principle, the suction could become so great that it would carry objects or people with it. We now want to analyse in detail whether the prerequisites for this are satisfied in the scene.

First of all, we need to consider how long it takes for this pressure equalization to occur. The calculation required for this is based on the behaviour of gases, where the gas considered here is simply air. How long the air flow to the outside lasts depends on three factors: the amount of air in the cabin, i.e., the volume of air with higher pressure, also the area through which the air flows out, i.e., the cross-sectional area of the window, and finally, the ratio between the external and internal pressures. The volume of a Lockheed Jetstar aircraft is 23.74 cubic meters. The cross-sectional area of the window is the product of its height and width and is therefore 0.16 square meters. The pressure ratio is about 3:1, which means that the internal air pressure is three times as high as the external pressure at 10,000 meters altitude. A calculation with these numbers shows that the suction will then last only about 0.8 seconds. In the scene, however, Goldfinger floats in the air for seven seconds before arriving at the window. One possible explanation for this would be that the suction due to the pressure difference is not solely responsible for Goldfinger's departure. Instead, the flight curve—Pussy Galore is flying almost vertically downwards in panic—could cause Goldfinger to move in the direction of the window. However, it is certainly not a parabolic flight, so the hovering motion of the crook and his subsequent motion toward the window could not be explained in this way. The best explanation for the duration of the scene being almost ten times too long is therefore the following: in order for spectators to enjoy Goldfinger's exit to the full, the scene is played in slow motion, at a much slower pace, in fact, at about a tenth of the real speed.

In addition to the duration of the pressure equalization, the next thing to investigate is whether the suction would really be strong enough to pull the powerful Auric Goldfinger over such a distance to the window. The famous Swiss physicist Daniel Bernoulli found the Bernoulli effect, named after him in the 18th century. This describes the relationship between the flow velocity and the pressure of liquids and gases. Furthermore, the so-called continuity equation must be taken into account. This means that for the flow in a pipe, the cross-sectional area of the pipe multiplied by the flow velocity must remain constant. The narrower the pipe, the faster the liquid or gas must flow through it.

During the event, a flow velocity of about 330 meters per second will prevail directly at the site of the hole in the window. This is actually the speed of sound, as explained further in the "Details for know-it-alls". The cross-section of the space parallel to the window has a size of eleven square meters, about seventy times the window area. The speed of the air in the room is thus 1/70 of the speed of sound, i.e., about 4.7 meters per second. To get a feel for this number, the Beaufort scale is used to classify the strength of wind and storms. Wind force 3 corresponds to a speed of about 5 meters per second. On the Beaufort scale, such conditions are described as a "weak breeze" in which thin branches and leaves begin to move and foam forms on the surface of the sea. Therefore, it's completely realistic to find in the movie scene that a book remains on a side table while its pages only flutter slightly. This would actually happen with a book in a breeze of wind force 3, but a handsome criminal like Auric Goldfinger would certainly not be set in motion. This confirms the theory that the suction is not responsible for his departure and that he falls out of the plane due to some other cause.

We've already exposed the translation error in the German version of the movie when Pussy Galore was aiming her gun at Bond on the plane and Bond is supposed to have said: *"The suction will take us both out into the space of the universe."* At an altitude of 10,000 meters, one would certainly not yet speak of the "space of the universe". Not until an altitude of at least 80 kilometers. However, if we assume for the sake of argument that Goldfinger's plane was actually rebuilt so that it could fly at this altitude, then both the duration of the suction and its strength would actually correspond to the values of the movie scene. But it would still be impossible to explain how James Bond survived all this without any damage to his person: the air in his lungs would also completely escape in this case. His lungs would be badly injured, but he would suffocate anyway because of the lack of oxygen. It is also strange that Goldfinger first gets stuck in the window, because this would close the leak and there would then be no further suction.

Let's just recap. The dangers of leakage in the pressurized cabin of a passenger aircraft are presented here much more dramatically than they actually are. The phenomenon of suction toward the outside known as "explosive decompression" nevertheless exists. If a passenger sits directly beside the window and the window is large enough for a person to pass easily through it, the suction would be enough to pull him out. But a passenger in the aisle seat would be quite safe, because the suction speed there would already be very low.

After Goldfinger has "taken his leave", the aircraft gets into heavy turbulence. The pilot Pussy Galore can no longer make an emergency landing, and she and Bond leave the crashing plane with a parachute at the last moment. They land on an island, waiting to be rescued. When Pussy Galore sees a helicopter she tries to draw attention to herself by waving. But the philanderer 007 pulls her to the ground with the words: *"Oh, no, you don't! This is no time to be rescued",* and both disappear tightly wrapped under the parachute …

Details for Know-It-Alls

When calculating the velocity of the air flow in the aircraft, the suction velocity at the window is of essential importance. When gases flow, the speed of sound is often the limit. For even faster flows, so-called supersonic or hypersonic flows, chaotic turbulence occurs, and this hinders any further increase in the flow velocity. Therefore the speed of sound, also called Mach-1, is the optimal case from the energy point of view. It is still a transonic flow, and the continuity equation remains applicable as a very good approximation. This equation is:

$$A_{\text{window}} \cdot v_{\text{window}} = A_{\text{plane}} \cdot v_{\text{plane}}$$

where A_{window} and A_{plane} are the cross-sectional areas of the window and the interior of the aircraft viewed side on, which are in a ratio of 1:70. As explained above, v_{window} is at most the speed of sound. This implies a speed of suction v_{plane} equal to a fraction 1/70 of the speed of sound, which is 330/70 = 4.7 m/s.

The duration of the suction is calculated by considering the mass throughput. We investigate the amount of air exchanged in a given time between two adjacent rooms with different air pressures, in our case the interior and exterior of the aircraft. Properties such as density, air flow speed, and specific heat play a role here. This can be summarized in the equation for mass throughput known as Fliegner's formula:

$$m / t = 0.04 \cdot A \cdot p_0 / T^{1/2}$$

where m/t (in kg/s) is the mass of gas flowing through the opening per unit time, $A = 0.16$ m^2 is the area of the opening, $p_0 \approx 90{,}000$ Pa is the pressure difference, and $T = 273$ K is the absolute temperature. With this equation one can then calculate the duration until the total mass $m = 28.5$ kg of air has flowed out of the aircraft. (Note that all quantities must be measured in the units used here for the formula to give the correct numerical value.)

4

The Physics in Moonraker

Hugo Drax: "*You have arrived at a propitious moment, coincident with your country's one contribution to Western civilization—afternoon tea. May I press you to a cucumber sandwich?*"
Bond: "*Thank you, no. Nothing at all.*"
Quote from the 11th Bond film *Moonraker*

In the final credits of the movie *The Spy Who Loved Me* from 1977, it was announced that James Bond would return with the movie *For Your Eyes Only*.[1] In view of the huge successes of the first *Star Wars* movie launched in the same year and the film *Close Encounters of the Third Kind*, however, these plans were changed and a space adventure was inserted. *Moonraker*[2] appeared on the screen in 1979 and is the eleventh film in the James Bond series. In contrast to *Goldfinger*, *Moonraker* simply took the title of Ian Fleming's book of the same name and devised a completely new plot for it.[3]

Moonraker was able to raise over 210 million dollars at the box office, with production costs of about 30 million dollars, and was thus the world's most successful film of 1979. It was the most financially profitable Bond film to

---

[1] The credits read literally: "James Bond will return in 'For your eyes only'". This film then actually followed *Moonraker in* 1981.

[2] The "moonraker" is in fact the highest propulsion sail on a fast cargo sailing ship, of the kind used in the middle of the 19th century.

[3] The novel *Moonraker* is the third novel in the Bond series written by Fleming in 1955. It is regarded as one of his best stories and is about the villain Sir Hugo Drax trying to extinguish London with a missile that carries a nuclear warhead. Parts of this story had already been used in the movie *The Spy Who Loved Me*.

© The Author(s) 2020
M. Tolan, J. Stolze, *Shaken, Not Stirred!*, Science and Fiction,
https://doi.org/10.1007/978-3-030-40109-2_4

date and was able to defend this title until the film *GoldenEye* in 1995. The inflation-adjusted box office result today is estimated at 720 million US dollars, which corresponds to 8th place among the 24 Bond films. 007 was played for the fourth time by Roger Moore, who died in 2017, and for whom Michael Lonsdale, playing the super villain Sir Hugo Drax, turns out to be an equal opponent. For many Bond fans, *Moonraker* is the second best movie in the series behind *Goldfinger*.[4]

But physics is not neglected in the film either. It begins spectacularly with an aerial battle, between Bond and a villain, which invites a more detailed analysis. Hugo Drax is planning the extinction of mankind. How could he do that? Is his space station really realistic, and could one simply fly there with a space glider, as shown in the film? The space station rotates and thus generates artificial gravity. Is that really possible? Lots of questions spring to mind and we shall now answer each one in turn.

How to Steal a Parachute

At the beginning of the film *Moonraker* a space glider disappears, and James Bond is sent to find it. In the opening sequence we find the pilot, the villain Jaws, and James Bond above the clouds in a plane threatened by a crash. The pilot who has disabled the plane tries to throw 007 overboard, but is catapulted out of the plane by a skillful blow from the secret agent himself. In the next moment James Bond is unexpectedly pushed out of the plane by Jaws, who also jumps a few seconds later. Since the pilot owns a parachute, James Bond tries to catch him by clever fall strategies in order to seize his parachute. After a certain time he succeeds in catching up with the pilot and removing his parachute in a spectacular aerial fight. The pilot is then left to fall mercilessly without a parachute. His presumably unpleasant fate is never revealed. Just when 007 has strapped on the parachute and considers himself to have reached safety, Jaws appears behind him at some distance. Since Jaws is clearly heavier than the secret agent, he catches up with him quite fast. However, the two have only a few brief moments of pleasure together, because James Bond pulls the rip cord of his parachute at precisely the moment when Jaws wants to ram his big iron teeth into Bond's leg—and thus shoots up in a flash. Jaws now also wants to open his parachute, but pulls so hard on his rip cord that it breaks off and the parachute can no longer open. At this point he makes out

---

[4] See, e.g., Jay Scott: *Moonraker: 007 in space as good as ever*. In: The Globe and Mail, June 30, 1979, p. 29.

a circus tent on the ground and tries to move towards it by rowing violently with his arms. He just makes it, dropping onto the tent roof and surviving the fall completely unharmed.

Here the following question arises: Would all that really work? Is it really possible for James Bond to catch the pilot in free fall? Why does Jaws catch up with him? And finally, how can the nasty Jaws survive a fall from such a great height onto the circus tent?

In this sequence, air resistance once again plays a decisive role. Since all bodies are accelerated equally by the Earth, all bodies fall equally fast,[5] and James Bond would have no hope of catching up with the pilot below him without air resistance or external propulsion. He falls out of the plane three seconds after the pilot and this lapse of time could not otherwise have been made up by any manoeuvre in the air—James Bond would always have remained three seconds behind the pilot. The same goes for Jaws. Despite his enormous weight, he can't catch up with James Bond without somehow exploiting air resistance.[6]

Some data concerning the protagonists are required for a detailed calculation. First there is James Bond, who is 1.83 meters tall, weighs 76 kilograms, and has an estimated but credible shoulder width of 45 centimeters. Jaws, on the other hand, is an impressive 2.20 meters tall, weighs 144 kilograms and has a shoulder width of about 55 centimeters.[7] Weight and size are also important for calculations concerning the pilot. His measurements can be roughly estimated from the film sequences. This results in a size of 1.86 meters, a weight of about 70 kilograms, and a shoulder width of 43 centimeters.

The total force acting on the falling body is composed of the downward weight force and the upward friction force caused by air resistance. While the weight force is simply the product of the mass and the acceleration due to gravity, the air friction is more complicated. We have already seen this for the Daniel Craig stunts and will repeat it again here. It depends on the body shape, more precisely on the streamlined shape, the air density, the cross-sectional area, and the speed of the falling body. The dependence on the speed

---

[5] This means that the same distance is covered in the same fall time. It doesn't matter which body falls. A feather would fall to the ground just as quickly as a hammer without air resistance. This was impressively demonstrated on the Moon by astronaut David R. Scott during the Apollo 15 mission in 1971. There is no air and consequently no air resistance, so that all bodies fall to the ground at the same speed.

[6] In the jumps that Daniel Craig made from the crane boom in *Casino Royale,* the corrections due to air resistance would have been minimal, as the fall times are very short. So nothing would change. Likewise in his fall from the train on the bridge into the water in the film *Skyfall,* air resistance doesn't matter much because of the short fall time. In contrast, air resistance plays the decisive role when James Bond catches up with the pilot!

[7] Jaws was played by the actor Richard Kiel. We have simply given his dimensions here.

is quadratic, i.e., doubling the speed corresponds to quadrupling the air resistance.

The streamlined shape of a body is described by the drag coefficient, known as the c_W *value*. This depends on the shape of the falling body. The highest c_W value is for a body hanging from a parachute. The c_W value of a downward-open half-shell, i.e., an open parachute, is quite large, with $c_W = 3.4$. For a smooth sphere, however, $c_W = 0.5$. Very streamlined bodies, such as the wings of an airplane, can have values as low as $c_W = 0.01$. Furthermore, the c_W value is greater when a falling person lies horizontally than when a person falls chaotically through the air, curled up and falling like a ball. These different c_W values were used as the basis for the calculations for this film scene.

In addition, the cross-sectional area of the body also plays an important role. This is the effective area contributing to air resistance. James Bond takes advantage of this fact to catch up with the pilot flying under him. The reduction of this effective area for air resistance can be clearly seen in Figure 4.1. This allows 007 to significantly increase his rate of fall.

In the calculations for this scene, we must take into account the fact that the density of the air decreases considerably with height. Air at sea level, for example, has a density of 1.2 kilograms per cubic meter, but at an altitude of 6000 meters above sea level, i.e., the height of the jump, this value drops to only 0.44 kilograms per cubic meter, a third of the value at sea level. The *Dynasys* program can be used to model and simulate dynamical systems in order to be able to calculate as accurately as possible with all the given data. *Dynasys* is a freeware, available on the internet. With this software the motions of bodies can be simulated, if all acting forces are given. The program then solves the associated equations of motion by numerical integration.[8] The times of each individual jump and the duration of the entire free fall are known from the film scene. Thus the initial conditions for the calculation are given. Then all times at which there was a change in any factor influencing the falling motions of the given individuals were determined by a precise analysis of the film scene. This refers to changes in the effective cross-sectional areas of the falling bodies and changes in the c_W values during the fall. In addition, there are the different masses of the three individuals, which of course do not change with time, as well as the c_W value of the parachute which James Bond opens at the end. A realistic value of twelve kilograms was assumed for the mass of the parachute. All parachutes are supposed identical. Thus the weight

---

[8] The program is ideal for use in the classroom because it reduces the amount of mathematics required for understanding to a minimum. It can be found on the Internet at http://www.hupfeld-software.de (as of March 2019).

$A_{Bond} \approx A_{Pilot}$

$v_{Bond} \approx v_{Pilot}$

A_{Bond}

$A_{Bond} < A_{Pilot}$

$v_{Bond} > v_{Pilot}$

A_{Bond}

Figure 4.1   In the upper picture Bond falls horizontally to the ground. His effective surface area A_{Bond} is relatively large and corresponds approximately to his height multiplied by his shoulder width. This area is approximately the same as that of the pilot A_{pilot}, i.e., the villain below him. That's why the two bodies would normally fall at about the same speed. The picture below shows how Bond reduces his effective air resistance area A_{Bond} in order to increase his speed of fall v_{Bond}

of twelve kilograms must be added to the weights of James Bond, Jaws, and the pilot. We also took into account the fact that the air density increases as an object falls further.

A simulation of the falling motion of the three individuals can be seen in Figure 4.2. The altitude at which each person is located is plotted as a function of the time that has elapsed since the pilot jumped.

The calculations show that the aircraft must have flown at least 6000 meters above sea level. Although this value is quite high, it can be explained by the fact that this jump was of course not planned. After all, the participants have to escape from a crashing plane. However, parachute jumps can still be carried out from a height of 5000 meters and more. The actual scene ends with Jaws hitting the circus tent after 135 seconds or 2:15 minutes. Figure 4.2 also reveals that James Bond flies 140 seconds further on the parachute before landing safely after a total of 275 seconds. The pilot probably hits the ground after a total fall time of roughly 205 seconds. Since this landing and James Bond's landing are not shown, we can only conclude this from the calculated data. Figure 4.2 also shows that all three individuals left the aircraft at different times. James Bond was pushed out of the plane by Jaws three seconds after the pilot jumped out. Exactly when Jaws himself jumped out is not shown in

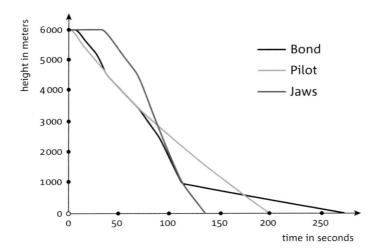

Figure 4.2 The height of fall of the protagonists is shown as a function of time. The vertical axis shows the height above the ground in meters with a starting height of 6000 meters. The horizontal axis shows the time that has elapsed since the pilot jumped. The pilot (light grey line) drops out of the plane at time zero and his altitude decreases. James Bond (black line) is pushed out by Jaws three seconds later. Only then does his height decrease with time. Jaws (grey line) jumps out of the plane 30 seconds after this scene. Therefore his altitude decreases only after a period of 33 seconds and remains constant at 6000 meters until then

the film and must therefore be reconstructed from the calculated trajectories. It turns out that Jaws must have begun his downward journey about 30 seconds after James Bond left the plane. The bend in James Bond's trajectory after about 28 seconds shows that he begins to fall faster, something he initiates by roughly halving his effective cross-sectional area and engaging in a steeper descent. Hence, 30 seconds after our top agent has jumped out, or after a drop of 1350 meters, he reaches the pilot at an altitude of 4650 meters above the ground. This can be seen from the coincidence of the dark line corresponding to Bond and the light grey line corresponding to the pilot in Figure 4.2. At this altitude, the two men begin their aerial fight for the parachute. At about 3450 meters, their paths separate again, and the pilot begins his deadly descent downwards without a parachute. At a height of about 2000 meters, Jaws has managed to position himself a little more than 100 meters from Bond due to his great weight. But he still has to fall almost 1000 meters more to catch the top agent, as 007 notices his enemy above him and both move in a dive with the optimal streamline shape. James Bond then needs only two seconds to release himself from Jaws by pulling on his rip cord. As a result, his c_W value and his effective cross-sectional area increase abruptly, leading to a sudden deceleration. So the secret agent can float safely to the ground, while Jaws continues to fall unbraked.

Figure 4.3 shows the rates of fall of the three individuals in meters per second as a function of the time elapsed since the pilot jumped. This graph makes it a little easier to read off the individual phases of the motion, as compared with Figure 4.2. The three individuals all start from the rest position, since the airplane is at a constant altitude of 6000 meters. The pilot quickly reaches a speed of fall of 54 meters per second, which corresponds to almost 200 kilometers per hour. As already mentioned, he can then slow down by bringing his body into an extended horizontal position and thus maximizing his effective cross-sectional area for air resistance. When the pilot meets James Bond and they begin to fight in mid-air, he again reaches a speed of 55 meters per second, which is his maximum speed. 007, on the other hand, reaches a top speed of 76 meters per second, or 274 kilometers per hour![9] After 30 seconds falling together, the pilot and Bond separate. After the air battle, the pilot has a speed of 118 kilometers per hour, which only slowly decreases until his unknown, but probably unpleasant end.

---

[9] For comparison, at 285 kilometers per hour, the top speed of the Porsche 911 Carrera is only slightly higher. However, James Bond would be much faster in free fall than the Aston Martin DB5 which he drives in several films, as it can only reach a top speed of 238 kilometers per hour.

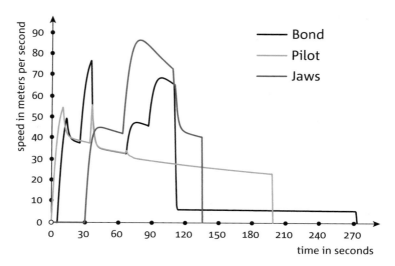

Figure 4.3 The rate of fall of the protagonists is shown here as a function of time. The vertical axis shows the speed in meters per second. The horizontal axis indicates the time elapsed since the pilot jumped. As long as the individual is still on the plane, their speed is zero. The dark line for James Bond, for example, begins three seconds after that of the pilot and 30 seconds before that of Jaws

Jaws, who at 144 kilograms weighs almost twice as much as James Bond, has even better prerequisites and can reach a significantly higher top speed. After about 80 seconds, he falls with a top speed of almost 86 meters per second, which corresponds to 310 kilometers per hour.[10] In the end, all three men fall at constant speed, with only James Bond moving slowly enough, at about 30 kilometers per hour, to be able to land safely. The pilot strikes the ground at about 100 kilometers per hour and would certainly not survive this.

Jaws, on the other hand, survives his crash completely intact, although his speed of fall is even greater than that of the pilot. However, his fall is also more gently slowed down by the circus tent. It has an estimated height of 20 meters, which is therefore available as braking distance. The deceleration time is about one second, as the exact analysis of the corresponding film section shows. At the time of his impact with the circus tent, Jaws will be falling at about 150 kilometers per hour. The force acting on him over the deceleration period of one second is then approximately 5700 newtons. This corresponds to about 4 g, i.e., four times his body weight.[11] For comparison, in a roller coaster ride, the passengers in a loop are also affected by up to 4 g, but over a shorter period

---

[10] This speed now corresponds roughly to the top speed of a Ferrari Testarossa, which at 390 H.P. can reach 320 kilometers per hour on the road.

[11] So-called "g-forces" will be dealt with in more detail in the next section.

of time. Therefore, even over a period of one second, 4 g still appears to be a tolerable load. This applies even more so to Jaws, who is clearly of an extremely robust nature.[12]

The opening sequence of the film *Moonraker* could therefore really occur just as viewers see it on screen if all the actors jump from an initial altitude of 6000 meters. It could theoretically have been filmed with a single jump. In reality, however, it was realized at a height of only 3000 meters and was composed of more than 90 individual scenes shot during 88 jumps.[13] The shooting of this scene lasting about two minutes took five weeks and was also relatively dangerous, because an ultra-compact parachute had to be developed, which could be worn under one's clothes without the viewer noticing it. With such a parachute it is not always sure that it will actually open when required.[14] But James Bond survives this initial adventure unharmed and the actual plot of the film can now begin.

Details for Know-It-Alls

In order to survive this precarious situation, air resistance is once again essential for James Bond. We have already seen in the Daniel Craig adventures that the top agent has to know a lot about air resistance, and we need here to investigate the dynamics of a falling movement even more closely. So let us consider still some more expert information regarding the free fall.

The first thing to clarify is why the Earth accelerates all bodies equally and why all bodies should fall to the ground at the same rate. This remained a riddle until the 17th century. Everyday experience shows that heavy bodies sometimes fall to the ground faster than light ones. After all, who would deny that a sheet of paper falls more slowly than a brick? However, if the sheet of paper is crumpled up, it falls to the ground much faster than before, although its mass has certainly not changed when it is crumpled up. The famous physicist Galileo Galilei already realized in the 17th century with the help of an impressive thought experiment that all bodies must be accelerated equally strongly by the Earth.

---

[12] Jaws survives a lot in the movie *Moonraker*. For example, he speeds into a concrete hut with a cable car, and the hut is completely destroyed, while he himself remains unharmed. Jaws also survives a crash over a high waterfall in a motorboat. An acceleration of 4 g over a period of one second is therefore not really a great challenge for such an indestructible character.

[13] This has been described in detail in the already quoted documentary *Best ever Bond* from the year 2002.

[14] The scene was also particularly dangerous for the cameraman. He had to jump off with a helmet camera weighing 3.5 kilograms and at the end open his parachute very slowly so that this weight would not break his neck.

Galilei gave a proof by contradiction.[15] The initial hypothesis is that heavy bodies are indeed accelerated more strongly than light ones. They would then have to fall faster, too. This assumption can now be contradicted by purely logical considerations without practical experiments. Assume first that we have a light body and a heavy body. If the light body is directly below the heavy body, then it should inhibit the faster falling motion of the heavy body, because it falls more slowly, according to the assumption.[16] In total, the two bodies together should fall to the ground faster than the light one, but more slowly than the heavy one. But if now these two bodies are connected into a single unit, then this unit is altogether heavier than the heavier of the two individual bodies. According to our initial hypothesis, this entire body would have to fall faster than the heavy single body. However, the same initial hypothesis already led to the conclusion that it had to fall more slowly overall. But it cannot fall faster and more slowly at the same time! So the initial hypothesis delivers a paradoxical result. It must therefore be false, and all bodies must be equally accelerated by the Earth—the proof by contradiction is complete.

The fact that all bodies are equally accelerated in the direction of the Earth's center does not contradict the observation that different objects fall at different speeds, since it is the *total* force acting on a body that determines its movement. In addition to the force of gravity, the second essential force is the air friction, which inhibits the falling movement, as we have already seen in the Daniel Craig scenes. This force is now different for different bodies of different sizes and ultimately ensures our everyday experience that heavy, streamlined bodies fall faster than light, extended bodies. Unfortunately, when air resistance is taken into consideration, the exact mathematical description of the falling motion is already very demanding. The corresponding equations of motion are so-called differential equations. Their solution requires higher mathematics, which we ourselves must withhold, even from the know-it-alls who are reading this section. With the program *Dynasys* it is, however, possible to set up these equations of motion and solve them by computer using a numerical calculation method.[17] This is how Figures 4.2 and 4.3 were produced. However, just from the simple balance of forces, it is possible to calculate the constant fall velocity, or terminal velocity, which occurs after a long

---

[15] Proof by contradiction is generally the most difficult thing logic has to offer. But at the same time it is also the most beautiful element of logic and always ensures a lasting aha experience.

[16] A cent piece alone also falls to the ground faster than a cent piece on a sheet of paper. Here the air resistance of the paper inhibits the motion. In our thought experiment, however, there is no air resistance, only different masses.

[17] For true experts, the program Dynasys uses a so-called Runge-Kutta method of 4th order to solve the relevant differential equations.

enough fall time, when air resistance is taken account. All the exact calculations carried out with the *Dynasys* program were additionally checked with this in mind. If F_{total} is the total force acting on a falling body, then

$$F_{total} = F_G - F_R$$

where the air resistance force F_R must be subtracted from the gravitation F_G because it counteracts it. The weight force is simply

$$F_G = m \cdot g$$

where m is the mass of the falling body and g is the acceleration due to gravity, with the exact value $g = 9.81$ m/s^2.

As before, for the air resistance force F_R, we have

$$F_R = \frac{1}{2} \cdot c_W \cdot A_{eff} \cdot \rho \cdot v^2$$

Here c_W is the so-called air resistance coefficient of the falling body, which depends exclusively on its shape, A_{eff} is its effective cross-sectional area, v is its instantaneous velocity and ρ is the density of the air, which varies with the height during fall. The quadratic dependence of v leads to the fact that the friction force, i.e., the air resistance, increases very rapidly with increasing speed of fall.

If the air resistance is equal to the weight force, then the total force acting on the falling body disappears and it moves on at a constant speed. This follows from Newton's first law. This constant velocity can thus be derived from the balance of forces: weight force = air resistance force, i.e., $F_G = F_R$. Putting the above formulas together and solving the resulting equation for v^2, we obtain

$$v^2 = 2 \cdot m \cdot g / \left(c_W \cdot A_{eff} \cdot \rho \right)$$

As the mass m is in the numerator of the fraction and the air resistance coefficient c_W and cross-sectional area A_{eff} are each in the denominator, this formula correctly reflects everyday experience: heavy streamlined bodies with small cross-sectional area fall faster than light non-streamlined objects with large area. James Bond exploits the quantities c_W and A_{eff} in the denominator of the above formula in his pursuit of the pilot through the air by reducing both values for himself and thus increasing his speed. Jaws, on the other hand, uses the fact that m is in the numerator of the above formula and his mass is almost twice that of our smart secret agent in order to catch up with the latter.

In general, the c_W value changes with the shape of the falling object and would have to be determined empirically for each body. For the calculations, however, the corresponding c_W values were estimated as well as possible by comparisons with similar objects. If a person falls more or less chaotically, then a value of $c_W = 1.5$ was used in the calculations with the program *Dynasys*. In the horizontally extended case, the person is not so streamlined. Therefore $c_W = 2$ was used for this case. Someone who falls head first is much more streamlined. This was taken into account by a smaller c_W value, viz., $c_W = 0.8$. Finally, the largest drag coefficient is $c_W = 3.4$ for an open parachute.

The dependence of the density of the air on the height was determined by the air pressure p. At altitude h, this can be calculated using the barometric altitude formula

$$p = p_0 \cdot e^{-h/h_0}$$

where p_0 is the atmospheric pressure at sea level and $h_0 = 8000$ m is the thickness of the atmosphere.[18] The air density ρ at altitude h is proportional to the air pressure p and is related to it as follows:

$$\rho = p / (R \cdot T)$$

where $R = 300$ J/(kg . K) is the gas constant and T the absolute temperature of the air in kelvin at height h. The constant value of 273.15 K or 0 °C for T was used to simplify the calculation of these motions.[19]

Inside the Space Flight Simulator

While the *Moonraker* space glider is being transported on the back of a Boeing 747, it is hijacked and the transport aircraft destroyed.[20] Bond receives an order from his boss M to investigate the company building the shuttle, Drax Industries in California, a little more closely. Here he sets out to find the

[18] This is not the real thickness of the atmosphere, but the altitude at which the density of the atmosphere is only 36.8 percent of the density at sea level.

[19] Here, of course, the altitude dependency of the temperature T should also be taken into account, since the air at high altitudes is significantly colder than at ground level. In order to keep the calculations manageable, however, this detail was dispensed with and the calculations were done using an average value. A more detailed analysis shows that this simplification has no significant impact on the results.

[20] The orbital gliders have the shape of a space shuttle. In fact, space shuttles were transported by NASA on the backs of converted Boeing 747 Jumbo Jets.

NASA scientist Dr. Goodhead, who will show him the shuttle production and parts of the astronaut training. 007 sees an attractive woman a few meters away:

Bond: "*Hmmm, hmmm!*"
The woman notices and approaches him.
Bond: "*Hello.*"
Woman: "*Good afternoon. Can I help you?*"
Bond: "*My name is Bond. James Bond. I'm looking for Dr. Goodhead.*"
Woman: "*You just found her.*"
Bond: "*A woman!*"
Woman: "*Your powers of observation do you credit, Mr. Bond.*"
Bond: "*James—to my friends.*"
Dr. Goodhead shakes the hand extended to her by 007 with the words: "*Holly Goodhead.*" (Figure 4.4)

To prepare his astronauts for the high accelerations during the rocket launch, Hugo Drax had a "space flight simulator" built.[21] This is a cabin located on a long arm. In a space flight simulator, people are accelerated by rotating the arm quickly to move them on a circular path. In this way one can

Figure 4.4 The MI6 agent James Bond (Roger Moore) and the CIA agent Dr. Holly Goodhead (Lois Chiles) work closely together in the film *Moonraker*

---

[21] Such centrifuge trainers are nothing unusual. For NASA astronauts, they are part of the standard training program for every space flight.

study the effect of forces on the human body. Holly Goodhead explains to James Bond how big these forces can become when they visit the space flight simulator, which is in fact a large centrifuge. They enter the room containing the device and we hear the following dialog:

Goodhead: "*This is the centrifuge trainer.*"
Bond: "*It simulates the gravity force you feel when shot into space.*"
Goodhead: "*The speed is controlled from up there.*"
Bond: "*Hmm.*"
Goodhead: "*Why not try it?*"
Bond: "*Why not?*"
James Bond enters the cabin and is wired and strapped down by Dr. Goodhead.
Goodhead: "*Strap yourself in firmly.*"
Bond: "*Yes, Doctor!*"
Goodhead "*Now your arms. That's to prevent you from knocking yourself out.*"
Bond: "*How fast does it go?*"
Goodhead: "*Up to 20 g, but that would be fatal. Three g is takeoff pressure. Most people pass out at seven.*"
Bond: "*You'd make a great saleswoman.*"
Goodhead: "*Don't worry. This is what we call a 'chicken switch'.*"
Holly Goodhead pushes a red button below the armrest.
Goodhead: "The moment the pressure gets too much, release the button and the power's cut off."
Bond: "*Just like that?*"
Goodhead: "*Come on! A seven year old can take three g's.*"
Bond: "*Trouble is, there's never a seven year old around when you need one.*"

So everything is clear. Bond says the centrifuge trainer "*simulates the gravity force*". This force can then be converted into multiples of g, the acceleration due to gravity, which is the first thing we should explain.

On the surface of the Earth, an attractive force acts on all bodies, ensuring that they do not simply float up into the air. This force of attraction is called gravity and is caused by the attraction between two masses, one of which is always the enormous mass of the Earth, while the other is the mass of the body.[22] This gravitational force now accelerates a falling body towards the center of the Earth. The acceleration toward the center of the Earth which the body experiences due to the mass attraction of the Earth is generally referred

---

[22] The mass of the Earth is an impressive $6 \cdot 10^{24}$ kg, that is 6 trillion times one trillion kilograms or 6 sextillion tons!

to as *g*. As already mentioned, it has the value 9.81 m/s², which is quite a large acceleration. A car that could apply 1 *g* would accelerate from zero to 100 kilometers per hour in 2.83 seconds. This is about the acceleration of a Formula 1 racing car. At 7 *g*, i.e., seven times the acceleration due to gravity, this would take only 0.4 seconds, and at 20 *g* only 0.14 seconds. 20 *g* is already an unimaginably large acceleration.

Because of Newton's second law, the effective force is then simply equal to the mass of the moving body multiplied by the acceleration. At an acceleration of 20 *g*, James Bond would therefore weigh about as much as a mid-range car – 76 · 20 = 1520 kilograms instead of his nominal 76 kilograms. The *g-factor* therefore simply indicates the factor by which one feels one's own body weight has increased. This effective force is therefore also known as the g-force. 20 *g* is far too much, because even with 7 *g* most humans become unconscious, as the charming Holly Goodhead correctly notes. Blood is forced into the legs, so that not enough oxygen can reach the brain. This initially leads to unconsciousness and can eventually mean death if it lasts too long or if the g-values are even higher. Only specially trained individuals in special suits, such as fighter pilots, can withstand a load of 13 *g* at short notice. Their suits are filled with a liquid whose consistency is similar to that of human blood. This fluid creates a counterpressure and prevents the blood from collecting in the legs. The 3 *g* Dr. Goodhead mentions for the launch of a space glider is also realistic. NASA considers about 3 to 4 *g* to be justifiable for its astronauts over a longer period such as a Space Shuttle launch.[23]

So James Bond is sitting in the centrifuge trainer waiting for it to start. Shortly after it sets in motion, Dr. Goodhead is called away and 007 is handed over to Chang, Hugo Drax's unpleasant henchman. The latter sits at the control desk of the space flight simulator, slowly increasing the speed and thus the acceleration forces acting on James Bond. A displayboard shows the numbers from 1 to 14. Here we can directly read off the g-values affecting the occupant. At 7 *g* on the display 007 already looks very worn out and wants to stop, but the *'chicken button'* operating the emergency brake does not work. It has apparently been sabotaged by Chang. Only at 13 *g* does our top agent, who is already hallucinating and sitting almost unconscious in the cabin, come up with a suitable idea. Q had given him one of his obligatory toys: arrows placed on his forearm, fired with small explosive charges when the hand is bent.

---

[23] According to the Guinness Book of Records, 179.8 is the highest ever measured g-force survived by a man, David Purley, in 1977. For an untrained person, on whom a g-force acts in the direction of the body axis, we may make the following observations: a g-force of 1–2 g is bearable without restriction, a narrowing of the visual field begins at 2–3 g, while there is already a tubular visual field and "greyout", a preliminary stage of "blackout", at 3–4 *g*. "Blackout" occurs from 4–5 g and unconsciousness in the range 5–6 *g*.

With a well aimed arrow shot on the control console, which Bond cannot operate directly because he is strapped in, he ends his supposed death ride. The top agent has been badly beaten, but has once again managed to pull his head out of the noose at the last moment. The rogue Chang, on the other hand, has disappointment written all over his face because he has failed to kill James Bond.

This scene will also be analyzed again in detail. Bond is certainly much tougher than normal people, who are already unconscious at 5–6 g, so he will already have chosen not to get out until 7 g. But the fact that he can really withstand 13 g over a period of several seconds already borders on a miracle. Here you can see how resilient he really is! However, the question now arises as to whether a force of 13 g is really being applied James Bond in the centrifuge trainer. This can be determined by an accurate analysis of the image of the cabin.

In the cabin where James Bond is seated, the weight force F_G acts vertically downwards and the centrifugal force Fz acts outwards. The centrifugal force is generated by the motion of the cabin along a circular path. It is the same force that pushes the occupants of a car to the side when driving around a sharp bend, or pushes the occupants into the seat when driving through a loop on the roller coaster.

The force F_{Bond} felt by James Bond is composed of his weight force and the centrifugal force acting on him, as shown in Figure 4.5. The resultant force can be determined by considering the other two forces as the edges of a rectangle and drawing the diagonal of that rectangle.[24] If you know the weight force and the centrifugal force, then the force effectively acting on James Bond can be calculated relatively easily, for example, using the Pythagorean theorem.[25] The angle of deflection of the cabin can also be read from Figure 4.5 and the centrifugal force acting on Bond can be calculated. The dependence of the centrifugal force on the deflection angle of the cabin is shown in Figure 4.6. From this, it can be deduced that, with an effective 13 g the car would have to be deflected by an angle of 85.5 degrees, i.e., it would have to be almost horizontal. When measuring the angle of inclination, however, it turns out that it is only 56.5 degrees, from which we can calculate backwards to deduce a centrifugal force of only 1.5 g acting on James Bond.

So something cannot be right here, especially since a centrifugal force of 1.5 g leads to an orbital time for the cabin of about 3.7 seconds. In the film,

---

[24] Mathematically speaking, this is simply a vector addition.

[25] Remember that, in a right-angled triangle, the sum of the squares of the sides at the right angle is equal to the square of the side opposite to that right angle: $a^2 + b^2 = c^2$.

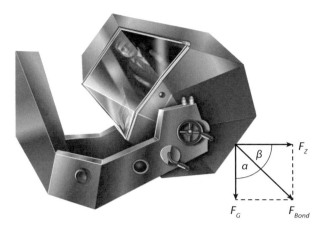

Figure 4.5 Forces acting on the cabin of the centrifuge trainer. Here F_G is the weight force and F_z is the centrifugal force. F_{Bond} is the resultant force felt by James Bond. The angle α describes the deflection of the cab from the vertical. For $\alpha = 0°$ the cabin is vertical

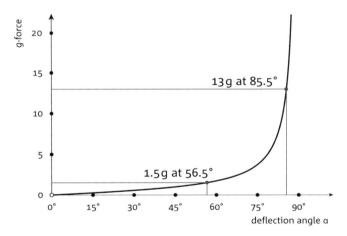

Figure 4.6 Centrifugal force acting on James Bond in units of acceleration due to gravity g (g-force) as a function of the deflection angle of the space flight simulator cabin (see Figure 4.5). An effective force of 13 g therefore corresponds to an inclination angle of the cabin of 85.5 degrees. However, the deflection measured in the film clip is only 56.5 degrees, which corresponds to a centrifugal force of only 1.5 g

however, the duration of the orbit is much shorter at 1.2 seconds and would fit better with the desired 13 g. All this can only be explained if the scene was actually recorded with an orbital duration of 3.7 seconds, and this would be why the observed inclination angle of the cabin is 56.5 degrees. Later, the scene was simply played at three times the speed, which fixes the orbit time,

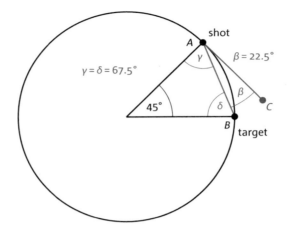

Figure 4.7 Illustration of the geometry of the arrow shot that saves James Bond's life in the space flight simulator. At point A, the arrow is triggered and would fly straight ahead to point C if the cabin were not rotating. The controls of the cabin that 007 wants to hit lie in this direction. Now the cabin moves during the flight of the arrow up to point B, so that the arrow would hit clearly to the left of the controls from Bond's point of view. He must aim at an angle β = 22.5 degrees to the right in order to hit his target

but of course does not affect the tilt angle. In this way, it is easy to demonstrate a little deception on the part of the James Bond producers ...

But how is it possible for James Bond to free himself from this deadly situation? He shoots an arrow at the controls of the cabin and brings the centrifuge trainer to a standstill. But his unerring shot is not as easy as it looks, because there are so-called pseudo-forces at work in the cabin.[26] These ensure that the projectile will not fly straight from 007's point of view, but will be deflected to the left (see Figure 4.7). This is due to the fact that the cabin continues to rotate during the flight time of the projectile—so the target is moving. In a straight shot, the projectile will travel from A to C and hit the controls of the cabin on the left. In order to compensate for this effect and hit the instruments opposite him dead on target, i.e., to shoot from A to B, James Bond must aim to the right at an angle of 22.5 degrees. This angle can be calculated from the time of a complete lap and the flight time of the projectile using the rule of three. This calculation would normally be very easy to do in one's head, but 007 has to do it under a load of 13 *g* in a flash just before he faints—under these conditions, it was a truly masterly performance.

---

[26] Pseudo-forces are so called because they only occur in a moving reference system. James Bond feels their effects in his cabin when he describes motions from his point of view. Seen from the outside, they do not exist, but their effects can be felt. Strictly speaking, the centrifugal force is also such a pseudo-force.

Details for Know-It-Alls

Figure 4.5 shows that the centrifugal force F_Z and gravity force F_G are at right angles to each other. The diagonal in this rectangle is the total force F_{Bond} acting on James Bond, which makes an angle α with gravity. A simple trigonometric relationship then provides the relationship[27]: $\tan\alpha = F_Z/F_G$.

These considerations are independent of the exact design and weight distribution of the cabin. They only depend on the position of the center of gravity with passenger. In the rest position, the center of gravity is just below the suspension axis and the cabin hangs upright. This is due to the vertical downward gravitation. When the centrifugal trainer rotates, the total resulting force F_{Bond}—the vectorial sum of the centrifugal force and gravity—takes the role previously played by gravity alone, and the cabin now "hangs" in the direction of F_{Bond}. It is therefore inclined at an angle α *to* the vertical. If the centrifugal force is to be a multiple of gravity, then the following applies: $F_Z = n \cdot F_G$, where n is then the g-factor. This results in the following formula: $\tan\alpha = n$.

This is shown in Figure 4.6. So for a force of 13 g acting on James Bond, $n = 13$, and the formula gives an inclination of $\alpha = 85.5°$. Conversely, we can calculate that the angle $\alpha = 56.5$ degrees only leads to $n = 1.5$.

To calculate the launch angle β for the arrow in Figure 4.7, we proceed as follows. A full orbit corresponds to 360 degrees, and estimating the radius of the centrifugal trainer to be $r = 5$ m, it takes about $T = 1.2$ s for an acceleration of 13 g, because the centrifugal force is calculated using the formula

$$F_Z = m \cdot r \cdot \omega^2$$

where $\omega = 2 \cdot \pi / T$. The force F_Z must then be equal to $n \cdot m \cdot g$, where in our case $n = 13$. With an estimated cabin length of about 1.5 m and a moderate launch speed of 10 m/s, the projectile will travel for exactly 0.15 s.[28] This corresponds to an angle of $360° \cdot 0.15/1.2 = 45°$ by which the cabin will continue to rotate during the shot. Since an isosceles triangle appears in the drawing in Figure 4.7 and the sum of all angles in the triangle is $180°$, the result is immediately $\gamma = \delta = 67.5°$. Hence, the required angle is $\beta = 90° - \gamma = 22.5°$. This is the angle at which James Bond has to aim to the right to hit the target directly opposite his cabin seat.

---

[27] The tangent (tan) of an angle is given in a right triangle by the quotient of the length of the side opposite the given angle and the length of the shorter side of the triangle adjacent to that angle.

[28] At a higher speed than 40 km/h, an arrow on the wrist could certainly not be fired without causing major injuries.

007 Has Magical Insight

After Bond barely survives the centrifuge trainer, Drax' attractive helicopter pilot Corinne Dufour shows him her boss's estate. It turns out that Hugo Drax is tremendously rich, as we can deduce from the following dialogue as they fly over a fairy-tale estate that looks like Vaux-le-Vicomte Castle, built in 1661 in Maincy near Melun in the Seine-et-Marne department of France:

Dufour: "*The Drax residence. Every stone brought from France. Cute, isn't it?*"
Bond: "*Magnificent. Why didn't he buy the Eiffel Tower too?*"
Dufour: "*He did. He was refused an export permit.*"

So Drax is really immensely rich, and Bond begins to investigate the property more closely. He beguiles the delightful Corinne Dufour and spends a night with her. While she is still asleep, he inspects a room, but initially finds nothing suspicious. Then suddenly Miss Dufour comes into the room:

Dufour: "*What are you doing?*"
Bond: "*You whetted my appetite for information. Is there a safe in here?*"
Corinne Dufour looks to the right at the wall where there is a chest of drawers.
Bond: "*Thank you.*"
Dufour: "*James. No!*"

007 opens the glass door in front of a clock, and a safe is pulled out of the chest of drawers. To open this safe, the top agent places a device that looks like a cigarette case on the door. This contains James Bond's cigarettes as well as a monitor. After switching on, a picture of the closing mechanism of the safe door appears. The image looks like an X-ray.[29] While James Bond turns the knobs to set the correct code, he can watch the closing mechanism work on the screen and soon manages to open the safe.

Dufour: "*That's amazing!*"
Bond (holding the device against Corinne's body): "*There you are. You have a heart of gold.*"
Dufour: "*18 carat.*"

---

[29] In principle, this could also be an ultrasound image. However, an ultrasonic signal breaks down when it hits air, and James Bond obviously does not use contact gel. What's more, the signal could not penetrate the air gap between the first metal layer, the safe door, and the actual closing mechanism.

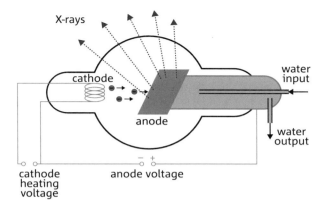

Figure 4.8 Schematic diagram of the most important components of an X-ray tube, all of which must be located in a vacuum bulb. The water serves to cool the anode

We shall now try to understand how James Bond's wonder device actually works, since apparently it can be used to look through massive 2.3 cm thick safe doors.[30] Only X-rays can penetrate thick steel plates. Therefore, the device in this scene must somehow be using X-rays. So let us first think about how this radiation can be produced. X-rays are generated in an evacuated tube in which a filament is heated, for example by a current flow, so that electrons escape. These electrons are accelerated by a high applied voltage, of the order of a few thousand volts, and then hit an anode made of copper. In the anode material they are rapidly slowed down, and it is this that produces the X-rays, as shown schematically in Figure 4.8.

Of the energy of the electrons hitting the anode, only a maximum of 1 percent is converted into X-rays. The remaining 99 percent is lost as heat, and increasing the temperature of the anode material. Therefore, X-ray sources must always be cooled. In principle, an X-ray tube can be relatively small and compact, but the question of whether it could be reduced to fit in a cigarette case is not even raised: the source must be inside the safe itself and illuminate the locking mechanism from there. An image like the one 007 sees is a typical X-ray image. His elegant case must therefore be an X-ray detector that measures the intensity of the incoming radiation. That's the only way it can work. We'll discuss this in more detail in a moment. Although the safe does not seem to be very large, it would not be a problem to integrate a compact X-ray tube and the corresponding high voltage supply.[31]

---

[30] The thickness of the safe door was determined by comparison with other objects.

[31] We prefer not to discuss whether the water cooling system also fits into the safe.

The X-ray radiation is now generated inside the safe. When it hits an obstacle, it will be absorbed, with a strength of absorption that depends on the nature of the material. Heavy materials such as lead, iron, or copper absorb more radiation than light materials such as water or carbon. Now let's turn to Bond's little gadget. Since the case is obviously a very small X-ray monitor, we have to assume that it is a modern semiconductor detector. Such a detector works just like an image sensor in a digital camera, in which the incident light is directly converted into an electrical pulse and further processed. Such sensors have also been developed in the last 20 years for the field of X-rays, but were not even a utopia in 1979, when *Moonraker* reached the cinemas. Once again, the British secret service seems to have been well ahead of its time!

To find out whether the safe door can really be X-rayed, we investigate how much the X-rays would be attenuated by the door. A measure of this is the material-specific attenuation coefficient, which is known for all substances. Figure 4.9 shows the percentage of X-ray radiation that would pass through a water or steel layer of a certain thickness.

The safe door has a thickness of 2.3 centimeters and is definitely made of solid steel. Figure 4.9 then shows[32] that the intensity of the X-rays arriving at the cigarette case will have dropped sharply to 1.5 thousandths of the initial value. This makes it clear that there must be an enormously powerful X-ray source behind the safe door, with an output of almost 100 kilowatts. This, however, results in further requirements. For example, such a strong X-ray source has to be cooled accordingly. Normally, such cooling is achieved by means of a water circuit. But since the X-ray source including the high voltage supply must also be in the safe, it seems quite difficult to place a water circuit there. But somehow, they must have succeeded.

So this is the only conceivable scenario to explain this movie scene logically. Nevertheless, one essential question remains unanswered: Why would Hugo Drax install an x-ray tube in his safe that would so conveniently help a secret agent to spy on him? Maybe this is just a precaution to support the rogue's weak memory. If he ever forgets the right combination, he will still be able—just like James Bond—to open the safe.

But meanwhile, Hugo Drax has noticed the betrayal by his helicopter pilot and has her killed by his hunting dogs. The material in Drax' safe takes Bond on to Venice, where he meets Holly Goodhead again, who reveals herself as a CIA agent. The American secret service also has Hugo Drax on the radar. At night, Bond enters a glassblowing shop and finds a secret laboratory. There,

---

[32] Strictly speaking, this can no longer be read from Figure 4.9, but must be calculated as described in the know-it-alls section.

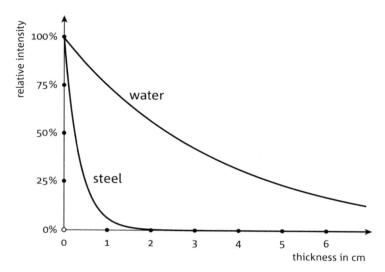

Figure 4.9 Typical intensity curve for X-ray radiation after passing through matter of a certain thickness. The intensity of the X-ray radiation decreases very quickly, exponentially to be precise, with increasing thickness of the material (shown in the x-direction). The curve for steel has already dropped to very small values, below 1%, after a thickness of only 2 centimeters

two employees fill containers with vials which have been carefully filled beforehand with some kind of liquid. Just as Bond is about to be discovered, a vial breaks and a gas is released. The laboratory is then automatically sealed behind Bond, and the two employees die very quickly from the toxic gas, while the laboratory rats survive. Just as in *Goldfinger*, it is surprising to see just how fast the gas spreads. In this laboratory, as in Goldfinger's room, there must be a very strong draught, although nothing else appears to be moving there. The top agent has also stolen a vial (Figure 4.10) which he hands over to M, before finally discovering that the poison containers are to be transported to Rio de Janeiro. The story continues there.

Details for Know-It-Alls

Like light, X-rays are electromagnetic waves, but with a wavelength a thousand times shorter and hence an energy a thousand times greater.[33] If the

---

[33] The energy E and the wavelength λ of electromagnetic radiation are related as follows: $E = h \cdot c / \lambda$. Here $c = 300,000$ km/s is the speed of light and $h = 6.62 \cdot 10^{-34}$ J·s is Planck's constant i.e. the quantum of action. The wavelength of visible light is 400–800 nanometers, the wavelength of X-rays is less than one nanometer. So it's about a thousand times smaller.

Figure 4.10 James Bond (Roger Moore) makes an interesting discovery in a secret laboratory

wavelength is very small compared to the dimensions of obstacles, then electromagnetic waves propagate in a straight line.[34] One speaks of light rays or X-rays. X-rays are very high-energy radiation. The more energetic the radiation, the more penetrating it is. This is the reason for the high penetration capacity of X-rays.

Figure 4.9 shows the typical intensity curve for X-rays that have passed through a certain thickness of a material. In fact, the figure shows the percentage of the X-ray radiation originally incident on the material that manages to penetrate through a thickness x. The relationship shown in Figure 4.9 is exponential, i.e., it can be described as follows:

$$I(x) = I_0 \cdot e^{-\mu \cdot x}$$

where I_0 is the initially irradiated X-ray intensity and $I(x)$ is the intensity measured after passing through a thickness x of the material. The material constant μ is called the attenuation coefficient and is a constant specific to the given material.[35] The symbol e in the formula is the Euler number e = 2.7182818... and serves to express the exponential function. Figure 4.9 shows the ratio $I(x)/I_0$ and thus the exponential decrease in the intensity with the thickness of the irradiated material. For the safe, whose door has a thick-

---

[34] Otherwise, so-called diffraction effects occur and the light no longer spreads in a straight line.

[35] It should be noted that the attenuation coefficient also depends on the energy of the X-rays.

ness of 2.3 cm of steel or iron, estimated from the dimensions of neighbouring objects, and using the attenuation coefficient μ = 2.8/cm for X-ray
radiation of energy 100 keV, the ratio obtained is

$$I / I_0 = e^{-6.52} = 0.0015.$$

Only about 0.15%, i.e., 1.5 thousandths of the X-rays penetrate through
the door of the safe.[36]

In order to operate a semiconductor detector and obtain an image with
suitable contrast, at least 30,000 X-ray pulses[37] per second and per pixel would
be required. If the cigarette case is to comprise such a detector and the pixel
size is set to 0.1 mm x 0.1 mm, then the total number of pixels of the detector
must be 252,000. It follows that 30,000 · 252,000 = 7.5 billion X-ray pulses
will be required per second on the detector. However, the radiation is attenuated by a factor of 0.0015 by the wall of the safe. The X-ray source must
therefore deliver 7.5 billion/0.0015 = 5 trillion X-ray pulses per second. The
energy of an X-ray pulse is E_{pulse} = 100 keV and thus the energy content of a
complete image will be E_{image} = 0.08 J.[38]

This energy must be made available as electrical energy from the X-ray
tube. For the electrical energy consumed over time t at a voltage U_R and with
a current I_R, we have the formula $E_{el} = U_R \cdot I_R \cdot t$.

James Bond sees a film on his detector in the cigarette case. The time for
each image can be assumed to be t = 0.04 s at 25 frames per second. At the
high voltage of U_R = 100,000 V, which would be necessary for the operation
of the X-ray tube, the value I_R = 20 µA follows for the required current, using
the relationship $E_{image} = E_{el} = U_R \cdot I_R \cdot t$. However, we must take into account
the fact that the efficiency of the X-ray tube is only 1%. This increases the
required current to I_R = 2 mA.

The power that the X-ray source must provide over the duration of the
18-second film scene is thus: P = $U_R \cdot I_R \cdot$ 18 s/0.04 s = 90 kW.

This value of almost 100 kW is so large that water cooling would definitely
be required. Reducing such a 90 kW X-ray source enough to fit into a small

---

[36] In fact, it will be much less, since 100 keV is already a very high energy for X-rays emitted from a tube.
So it really must have been a very strong source built into the safe.

[37] It turns out that the energy of X-rays is transferred to matter in small portions. These portions are called
X-ray quanta. Each X-ray quantum has an energy of $h \cdot c / \lambda$. The X-ray pulses mentioned in the text
actually refer to these X-ray quanta. Max Planck used the quantity h for the first time in 1900 to explain
the thermal radiation spectrum of a so-called black body. That's how he founded quantum theory. For
this achievement he was awarded the Nobel Prize for Physics in 1918.

[38] For experts, one electronvolt (eV) is the energy $1.6 \cdot 10^{-19}$ J. Thus 100 keV corresponds to an energy of
$1.6 \cdot 10^{-14}$ J. This is a typical energy for X-rays that would penetrate metals.

safe is an astonishing technical achievement—but we will see that Hugo Drax has much more to offer.

How Does "the Biter" Bite?

In Rio, Bond meets Holly Goodhead again on the viewing platform of the Sugarloaf Mountain. They discover that Drax Corporation cargo planes are taking off from the airport. The supposed villain seems to be planning something. Meanwhile he has hired Jaws as a new assistant. When 007 and Holly are in the cabin of a cable car, Jaws sabotages their further ascent by cutting a thick steel cable smoothly with his metal teeth. This explains why Jaws is also called "the biter". The cabin comes to a standstill at a lofty height, and Jaws is getting closer to them all the time. After Bond has convinced himself that this gentleman cannot be harmed with blows, he and Holly manage to escape. They slide along a chain looped over the steel cable towards the valley station and jump off in time, while behind them Jaw's gondola crashes into the building.

Now we would like to analyze how Jaws manages to bite through the thick steel cable with his teeth. First of all, it should be noted that a tooth could never bite through a steel cable. At 15 kilograms per square millimeter, the hardness of enamel is only about half that of steel. Therefore one could never cut a steel cable with normal teeth, because only a harder material can cut a softer one. But as you can see in the movie, Jaws has metal teeth and would be able to cut a steel rope made of inferior material. However, humans can only apply a biting pressure of 12.5 kilograms per square centimeter and thus a biting force of 800 newtons.[39] The maximum biting force an animal can exert on the Earth is 18,000 newtons, and this is reached by the white shark. So if, for fun's sake, we assume that Jaws can achieve exactly this force, could he then cut a steel rope with his teeth, as shown in the film?

This question was investigated experimentally by the *MythBusters*[40] television team. There, a set of sharp teeth made of specially hardened steel was constructed and an attempt was made to cut a steel cable about one centimeter thick. It turned out that even a biting force of 100,000 newtons was not sufficient to cut the rope, whence even more than five times the power of a white shark would not be enough to realize the film scene. So here

---

[39] After all, this corresponds to a weight of 80 kilograms!

[40] ¸The myth of Jaws' teeth was put to the test in the 98th episode, which was broadcast in 2008.

we have to resign ourselves to the fact that this is just not feasible, with the best will in the world, and making the most creative assumptions[41]!

After Bond and Holly have survived the adventure with the cable car, the CIA agent is kidnapped by Drax' henchmen. Bond, on the other hand, manages to flee and reaches MI6's Brazilian training camp, where Q presents the analysis results of the contents of the vial. It is the extract of a plant that only kills humans, but has no effect on other animals or plants. M instructs Bond to have a look around the lower Amazon region from which the plant originates.

Off into Space!

007 finally finds Drax's secret base, where Holly is being held. He is also captured, but immediately manages to free himself and Holly. Drax has his Moonraker space shuttles launched one after the other. Bond and Ms Goodhead succeed in entering the last one dressed as pilots and are launched into space with an unknown target.

Apparently, the Moonrakers from Drax are nothing more than space shuttles.[42] These were manned spacecraft developed by NASA in the 1970s, then launched regularly into space from 1981 until they were decommissioned in 2011.[43] They consisted of the orbiter, which looks like an airplane and was able to return to Earth, an external giant fuel tank, and two smaller fuel rockets which were blown off immediately after take-off and also reused.[44] NASA carried out a total of 135 flights, two of them resulting in fatal accidents in which each of the seven crew members died.[45] A total of five orbiters capable of space flight were built.[46]

---

[41] In the film, a liquorice rope was used instead of a steel cable, and this of course makes things much easier for Jaws.

[42] The last flight to the moon, Apollo 17, took place in 1972. Afterwards the Apollo program was terminated and a systematic space program was only taken up again by NASA in 1981, with the Space Shuttles.

[43] The biggest problem with a journey into space cannot be expressed more accurately than by a quotation attributed to Wernher von Braun: *"Two problems have to be solved in the conquest of space: gravity and paperwork. We could have handled gravity."*

[44] The development of the Space Shuttle originally aimed at significantly reducing the costs of a space flight by using reusable rocket stages. This turned out to be a gigantic fallacy, at about 500 million dollars per launch.

[45] These were the Challenger and Columbia catastrophes in 1986 and 2003, resulting in an astronaut death rate of 1.5 percent. By way of comparison, the probability of being killed in an aircraft crash is less than one hundred thousandth of a percent!

[46] There is another non-space glider involved in atmospheric flight tests. As the successor to the Space Shuttle, the Orion spacecraft is under development, which after two unmanned test flights in 2014 and 2018 will also be manned in the coming years.

The spacecraft was able to carry 24.5 tons of payload and up to eight astronauts into a low Earth orbit, at altitudes between 200 and 650 kilometers. With a total mass of 2000 tons of shuttle and filled tanks, this implies a ratio of 80:1. For each ton of payload, 80 tons of fuel must be consumed. Rocket engines are always inefficient in this sense. This is a consequence of the rocket equation, which itself is based on fundamental physical principles. There will never be a small light rocket that can fly into space. Moreover, with the help of docking adapters, the shuttle was able to dock to a space station such as the Russian Mir or the International Space Station ISS. This ability to transport crew and freight simultaneously made the shuttle very versatile. The construction and supply of the Mir and ISS space stations were central components of the shuttle missions.[47]

The film *Moonraker* came to the cinemas in 1979. At that time it was already clear that NASA would build a reusable space glider. The striking shape with the orbiter in the middle and the large central tank as well as the two small booster rockets on the side had already been clear since 1972. This design was apparently adopted by the *Moonraker's* makers, because Drax's spaceships look just like space shuttles. Drax must, however, have had a much better infrastructure than NASA. He maintained a fleet of six shuttles, which he could apparently send into orbit at the same time. NASA, on the other hand, has only ever had one space shuttle in Earth orbit at any given time.

What's strange is the scene when the large central fuel tank burns out. The empty tank is blown off just like the "real" Space Shuttle and then falls down. However, during the manoeuvre, the Earth is above the space glider in the picture, so that the tank should logically fall upwards. The tank must therefore have its own engine and get shot far away. Apparently, Hugo Drax was aware of the growing problem of space debris which orbits the near-Earth area and thus poses an ever-increasing threat to satellites.

Details for Know-It-Alls

Let us now explain rocket propulsion in a little more detail to understand why space shuttles and other space rockets always need so much fuel and therefore have to be so large. From Newton's laws of mechanics, one can derive the rocket equation, which is the basis of every form of recoil propulsion. It tells us the velocity v of a rocket or spacecraft which initially has velocity v_A and

---

[47] Since 2011, the ISS has only been accessible to Russian Soyuz capsules and space ships from SpaceX.

mass m_A and ends up with mass m_E, after ejection of its fuel, with the fuel being ejected with velocity v_F relative to the rocket. It reads[48]:

$$v = v_A + v_F \cdot \ln\left(m_A / m_E\right)$$

where "ln" stands for the natural logarithm, i.e., the logarithm to the base e.[49] The equation is valid in space, i.e., without air friction and without gravity. For a rocket to be launched on Earth, appropriate modifications would have to be made, but we can omit them here because we are only interested in the principle. It can be seen that the velocity v_F at which the fuel is ejected has a direct influence on the velocity v *of* the rocket. Doubling v_F results in doubling the rocket velocity.

But the really bad thing for space travel is the logarithm in the rocket equation. This logarithm is the most depressing thing about all space travel, because it ensures that it will never be "easy" to fly into space. Why is this? m_A/m_E is the ratio of the initial mass of the spacecraft including the fuel to the final mass, i.e., to the mass without the fuel, which has been ejected at the back. For example, if the mass at the beginning is twice as large as at the end, i.e., one starts with exactly the same weight of fuel as the spaceship itself weighs, and $m_A/m_E = 2$. However, the term in the rocket equation is $\ln(m_A/m_E)$, so the result for the velocity is not $v = v_F \cdot 2$, but only $v = v_F \cdot \ln(2) = v_F \cdot 0.69$. In the best case, $m_A/m_E = 6$ can be achieved. Then $v = v_F \cdot \ln(6) = v_F \cdot 1.79$ and not $v = v_F \cdot 6$. The logarithm is the reason why rockets have to be so large, because one needs a relatively large ratio m_A/m_E to reach the corresponding velocities.[50] If, for example, one wants to reach the velocity of $v = 11.2$ km/s, which is the escape velocity from Earth,[51] with a rocket that uses one of today's fuels, ejected at a maximum of $v_F = 5$ km/s, it follows from the rocket equation that a mass ratio of approximately $m_A/m_E = 9.4$ will be required. But there is no way of realizing that technically. A ratio of 6 is the best that can be achieved. In other words, one can't even overcome the Earth's gravitational pull.

---

[48] The rocket equation was first established in 1903 by Konstantin Eduardovich Tsiolkovsky and is therefore often referred to as the Tsiolkovsky equation. It only applies to velocities that are much lower than the speed of light, but this is no restriction here, because even 100,000 km/h is still much slower than the speed of light.

[49] e is the important Euler number in mathematics: $e = 2.71818 \ldots$

[50] If we wish to double the speed of a rocket, we have to eject the factor $e^2 \approx 7.4$ more mass at the rear. Four times the speed requires a factor $e^4 \approx 54.6$. This is a consequence of the logarithm in the rocket equation.

[51] This is the speed at which one can completely escape from the Earth's gravitational field.

This is why, in order to achieve this, a multi-stage principle must be applied, i.e., the rocket equation must be applied several times in succession, as shown in the following example. We assume a two-stage rocket, whose stages have masses of 100 tonnes and 20 tonnes, respectively, with each consisting of 90 percent fuel, i.e., structural masses of 10 tonnes and 2 tonnes, respectively.[52] The payload is also 2 tonnes. The basic rocket equation is applied twice, adding the contributions of the two stages, since the final velocity after burning off the first stage corresponds to the initial velocity at the ignition of the second stage. It then follows that

$$v = v_F \cdot \ln\left(100 + 20 + 2 / 10 + 20 + 2\right) + v_F \cdot \ln\left(20 + 2 / 2 + 2\right)$$
$$= v_F \cdot 1.34 + v_F \cdot 1.70 = v_F \cdot 3.04$$

For comparison, the result for a single-stage rocket with the same fuel and structural mass would be

$$v = v_F \cdot \ln\left(100 + 20 + 2 / 10 + 2 + 2\right) = v_F \cdot 2.16$$

If we now assume v_F = 5 km/s, then we see that the single-stage rocket with v = 10.8 km/s does not reach the escape velocity of 11.2 km/s, while the two-stage rocket with v = 15.2 km/s is clearly well above this velocity. So the only way to say goodbye to the Earth's gravitational field is by using multi-stage missiles.[53] We can see again how frustrating the rocket equation actually is, because multi-stage missiles are technically much more complex than single-stage ones, and it is not possible to put an arbitrary number of stages on top of each other.

The space shuttle was sent into space with a single-stage rocket so it was not possible to leave the Earth's gravitational field with this vehicle. Since the Moonraker space gliders were apparently nothing more than space shuttles, it is clear that they could not move far from Earth either. Therefore, the plot of the film surely takes place in a near-Earth orbit and not far away, near the Moon.

---

[52] Ninety percent fuel would then be a ratio of 1 to 10, which is already above the best technically feasible ratio of about 1 to 6. The structural mass refers to the mass of the rocket without fuel and without payload.

[53] The Saturn V moon rocket had three stages. Rockets that transport space probes like the Pluto probe *New Horizons* into space to explore the Solar System often even have four stages.

Weightlessness and Artificial Gravity

Along with the other space shuttles, Bond and Holly Goodhead head for Drax's secret space station, which apparently can't be seen from Earth. The audience witnesses one of Drax's spaceships docking at the station and the villain's henchman hovering into the station. Obviously, there is weightlessness, as is also the case inside the ISS. The gangster moves to a desk and flips a few switches. The almost circular space station is immediately set in rotation by the ejection of fuel. As a result, an artificial force of gravity is apparently generated, corresponding to 80 percent of the gravity prevailing on Earth, as a measuring device on the console indicates. This artificial gravity makes it possible for everyone to stay in the space station for a longer time and to move freely there without any problems.

It is known that in weightless conditions the muscles lose mass, and after a long stay in space, astronauts can no longer walk independently when they return to Earth. We can therefore conclude from this that this space station was built to allow people to stay there for long periods of time. Otherwise there would have been no need to attach importance to artificial gravity. But does artificial gravity really work?

The space station is rotated by the emission of fuel. As with the centrifugal trainer, this generates a g-force. But man has no sense organ for the kind of force that acts on him. A correspondingly large centrifugal force that presses the occupants of the space station against the wall can thus be used as artificial gravity. The wall would then play the role of the floor that attracts you. In the film a measuring instrument shows $0.8\ g$. After the rotation is switched on, an artificial gravity of 80 percent of the Earth's gravity thus prevails in the space station. But is this realistic, and how much fuel would be needed to produce this artificial gravity?

From the given gravity of $0.8\ g$ the radius of the space station can be calculated. This requires the rotation speed of the station. It is determined directly by careful analysis of the film images. If we superimpose the first and the last image of the section in which the space station is completely visible, the rotation angle it has covered in a certain time can be measured. From the time of 2 seconds that the space station needs to continue turning these 5 degrees, the time for a complete revolution can be calculated immediately to be 2 . 360 degrees/5 degrees = 144 seconds. The size of the centrifugal force depends directly on the radius of the space station. Since the strength of the artificial gravitation of $0.8\ g$ is known, the required radius of the space station can be calculated. For a rotation speed of one revolution every 144 seconds, this yields a radius of 4.1 kilometers. By way of comparison, the International Space Station (ISS) has a length of 98 meters and a span of 109 meters.[54] Even a megalomaniac super villain like Hugo Drax could hardly build such a huge space station.

But the actual radius is only about 130 meters, as can be seen from a comparison with the size of the docked space gliders, which are nothing more than space shuttles, as we saw in the previous section. If the artificial gravitation prevailing at the station is calculated from this radius and the measured rotation time of 144 seconds, then a value of only 2 percent of the Earth's gravitation, i.e., 0.02 *g,* *is* obtained. Conditions in the space station are almost weightless—so it can't work that way!

The error here is probably in the rotation time of the space station, which would have to be only 25.5 seconds instead of the 144 seconds determined on the basis of the images, in order to achieve an artificial gravitation of 0.8 *g.* Perhaps the film scene used to measure the orbit time was played six times too slowly for dramaturgical reasons.[55]

The amount of fuel required for the rotation of the space station can be determined from conservation of momentum. The momentum of an object is the product of its mass and its velocity. The direction of the velocity is important. Conservation of momentum can be illustrated by the example of a person standing on skates holding a ball. As long as the person stands still, the total momentum of both is naturally zero. If she throws the ball away, the ball and the person have equal and opposite momenta, which still have to add up to zero because of conservation of momentum. One momentum is positive and the other negative. This only works if the two velocities point in opposite directions, i.e., the ball quickly flies forward and the person must therefore slowly glide backwards, because of her greater mass.

The same principle can now be used to determine the amount of fuel required. We assume that the momentum of the escaping fuel is completely transferred to the space station. Fuels can escape from a rocket at a speed of about 3000 meters per second. Since Hugo Drax's space station is about the same size as the ISS, its mass can be estimated at 400 tonnes. With the help of momentum conservation, we can now calculate that about four tons of fuel are needed to generate 0.8 *g* in the space station.[56] This corresponds to about

---

[54] If "size" of the space station means the area it covers in the sky, then Hugo Drax's circular space station would be about 1600 times bigger than the ISS!

[55] Or it's a simple flaw in the film. The production company would never have expected anyone to actually calculate this!

[56] However, these four tons must be ejected from two nozzles pointing in opposite directions in order to set the station in a stable rotation. Otherwise an unwanted linear motion is superimposed.

one percent of the total mass of the space station. This is the main reason why this method has not yet been used to generate artificial gravity on board the ISS, or earlier, on board the Russian Mir space station. This enormous amount of fuel would have to be transported at great expense into the appropriate Earth orbit, and this would be far too costly.

A further problem would be the pseudo-forces generated by the rotation, as with the space flight simulator. As long as you don't move, everything will be fine inside the space station. However, since the centrifugal force depends on the radius, the space station would have to have a very large radius or a high rotational speed to ensure that there is sufficient artificial gravitational force and the people on board do not feel too large differences in that gravitational force between different areas of the space station. Furthermore, the same effect that caused James Bond in the centrifuge trainer to aim his arrow 22.5 degrees too far to the right would have to be constantly taken into account in the space station. With each action, it would have to be kept in mind that any target would also continue to rotate during one's own motion. Effectively, a resident of the space station would have the feeling that someone was pulling him to the side with every movement.[57]

So production of artificial gravity by rotation of a space station is quite realistic and could really work. In practice, however, such experiments are not carried out because the additional quantities of fuel that would be required for rotation are then saved. Furthermore, two other facts about Drax's space station are a little strange. First, approximately 100 Space Shuttle flights were needed to complete the ISS project. This would also apply to the perhaps even larger space station of Hugo Drax. Has Hugo Drax managed to launch a space shuttle a hundred times from his launch site in the Brazilian jungle without the British or American secret service noticing? Secondly, the film says that the space station cannot be located from Earth because of a "radar camouflage shield". Although stealth technology, which allows an aircraft to be camouflaged from radar beams, has been in existence for quite some time, the space station would be easy to discover in its orbit around the Earth, using a normal relatively small ground-based telescope, without any radar beams at all. In the movie you can see that the space station isn't too far away from Earth. It is probably about 300 kilometers above the Earth's surface and thus at a similar altitude to the ISS. Since a secret service will always search the sky intensively for spy satellites, it is a miracle that the British MI6 and American CIA never noticed this large construction in the sky, just like the many rocket launches.

---

[57] The physicist speaks here of the fact that the so-called Coriolis force would be at work. The Coriolis force, like the centrifugal force, is a pseudo-force that does not occur when viewed from the outside.

Entire departments must have been looking the other way. These secret services don't seem to be working that well.

A total of six Moonrakers have brought many people to the space station, always very attractive young couples. They are Drax' employees, to whom he explains his diabolical plan while standing on a platform:

Drax: "*First there was a dream. Now there is reality. Here in the untainted cradle of the heavens will be created a new super-race, a race of perfect physical specimens. You have been selected as its progenitors—like gods. Your offspring will return to Earth and shape it in their image. You have all served in humble capacities in my terrestrial empire. Your seed, like yourselves, will pay deference to the ultimate dynasty which I alone have created. From their first day on Earth they will be able to look up and know that there is law and order in the heavens.*"

Bond and Holly Goodhead also observe this spectacle. Drax plans to use his poison gas to exterminate mankind completely. Then he wants to rebuild humanity with himself at the top using the personnel in his space station, which according to him consists exclusively of mentally and physically outstanding people. No Bond villain has ever come up with a bigger and more diabolical plan, neither before nor after!

007 cannot prevent Drax from firing three globes of poison gas towards the ground. They all contain so much poison that millions of people will die if they reached Earth. Bond then succeeds in pressing a large red button with the inscription "*Do not use unless station secured—Emergency stop*". This stops the rotation of the space station and turns off the radar cloaking device so that it will now be visible from Earth. Gravity disappears and chaos breaks out immediately. The Americans then launch a space shuttle with a unit of the US Marines on board, and it comes to a showdown in space.

Details for Know-It-Alls

First we want to explain in more detail what "weightlessness" is. Contrary to popular belief, weightlessness is not the absence of gravity or gravitation. The International Space Station (ISS) is located about 400 km above the Earth's surface. Since the Earth's radius is over 6000 km, it is quite clear that gravity must still be present at this altitude. A more precise calculation shows that about 90% of the acceleration due to gravity, i.e., 0.9 g, still act there. So why are the astronauts in the ISS weightless?

The effect of gravity can be seen, for example, when an apple falls from a tree to the ground or when we are pressed to the ground by our own weight. What we feel as "gravity" and usually equate with gravity is due to the fact that the ground exerts a counterforce from below, as stipulated by Newton's third law. However, this does not apply uniformly on all parts of our body, but most strongly on our feet, which must carry our full weight, and much more weakly on the neck, which only carries our head. That's why our body gets a little compressed. Weightlessness is now the absence of gravity, that is, the force that the ground exerts on us. This is only possible when a body is in free fall, so only a freely falling body experiences weightlessness. But that would mean that an astronaut in the ISS must be constantly falling to Earth. How can that be possible?

To understand this, let us consider a thought experiment which Newton himself conceived of and which is illustrated in Figure 4.11.

The figure shows that the speed of the astronaut is so high that he permanently falls around the Earth. Actually, he does fall freely downward, but the Earth is no longer there when he arrives. So an astronaut is in free fall all the time, without ever hitting the Earth. This is the reason for the weightlessness in the ISS and it's the same on board Drax's space station until the artificial gravity is activated by the centrifugal force.

For centrifugal forces, the following expression holds:

$$F_Z = m \cdot r \cdot \omega^2$$

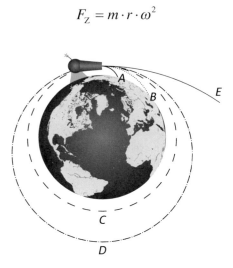

Figure 4.11   Newton's thought experiment of a horizontally launched cannonball. It reaches the ground at point A. If we increase the speed, it will land at point B. If the initial horizontal speed is sufficiently high, it will finally circle the Earth weightlessly (orbits C and D). At very high speed, it can even leaves the Earth's sphere of gravitational attraction (path E)

where $\omega = 2\pi/T$ is the angular frequency expressed in terms of the orbital period T, while m is the mass of the rotating body and r is its distance from the center of rotation. This force is therefore always large for large distances and short periods. Since the centrifugal force is supposed to act as artificial gravity, it must correspond in the space station to the weight force $m \cdot g$ of a body of mass m. However, in the space station only 0.8 g are effective, so the effective artificial weight force is $0.8 \cdot m \cdot g$. The above formula then gives

$$m \cdot r \cdot (2\pi / T)^2 = 0.8 \cdot m \cdot g$$

From this we can deduce the radius r of the space station:

$$r = 0.2 \cdot g \cdot T^2 / \pi^2,$$

which implies the unrealistically large radius of $r = 4100$ m when we insert the known numbers and the actually measured orbital period $T = 144$ s. Conversely, the artificial gravity can be calculated by transforming the above equation for a known radius, replacing the number 0.8 by n and solving the equation for n. If one does this and enters the realistic value of $r = 130$ m into the formula, the result is $n = 0.02$, implying an artificial gravitation of 0.02 g in the space station.

The amount of fuel required to rotate the space station is determined by conservation of momentum. The momentum is the product of mass and speed and must be equal before and after the fuel is used. Hence,

$$M_{\text{station}} \cdot v_{\text{station}} = m_{\text{fuel}} \cdot v_{\text{fuel}}$$

where m_{station} and v_{station} are the mass and speed of the station and m_{fuel} and v_{fuel} are the mass and speed of the ejected fuel. This formula can be rearranged to find m_{fuel}. Since $v_{\text{station}} = r \cdot 2\pi/T$, we obtain $m_{\text{fuel}} = 2\pi \cdot m_{\text{station}} \cdot r/ (T \cdot v_{\text{fuel}}) = 4270$ kg.

We have used $m_{\text{station}} = 400$ tons and $r = 130$ m for the mass and radius of the space station, and we have assumed that the fuel is ejected at a speed of $v_{\text{fuel}} = 3000$ m/s and inserted $T = 25.5$ s for the orbital period. As can be seen, these figures imply that a fuel mass of about four tonnes would be required to achieve rotation. Experts will have noticed that strictly speaking we should not have argued in terms of the conservation of momentum, but rather the conservation of angular momentum, since what we have here is a rotational movement. Thus, the angular momentum would be the physical quantity that matters, and which should be preserved overall. In this special case, however, such a consideration would lead to the same result.[58] On the other hand, it is clear that the four tonnes of fuel must be ejected from two oppositely directed nozzles for the station to rotate. Otherwise the conservation of momentum would also lead to a linear movement of the whole station.

Destructive Laser Beams

After extensive fighting between the crew of the station and the Marines, Bond shoots the villain Drax with an arrow from his wrist and transports him with the words, "*Heartbroken, Mr. Drax. Allow me. Take a giant step for mankind.*" into space. The station breaks apart, but 007 and Holly Goodhead are just able escape in a Moonraker with the help of Jaws. They then track the three globes with the toxic substances that are destined to kill millions of people when they reach Earth. Bond activates the laser built into the glider and tracks the death capsules.

Holly: "*I have the three globes on the screen. We should have a visual in a minute.*"
Bond: "*There it is.*"
Holly: "*The laser's on automatic. Code 945 set.*"
Bond enters the code with a keyboard and destroys the first globe with a pre-
    cise shot.
Bond: "*One down, two to go. Number two straight ahead.*"
The orbiter begins to vibrate.
Bond: "*What is it?*"
Holly: "*We're skipping on Earth's atmosphere.*"
Bond: "*In range.*"
007 destroys the second globe, and Holly Goodhead races after the third.
Holly: "*We should see the last one soon.*"
Bond: "*There it is.*"
The space glider enters the outer layers of the atmosphere and its tip becomes
    red-hot.
Bond: "*It's getting hot.*"

---

[58] Under the present circumstances, the angular momentum is simply the product of the momentum and the distance r to the axis of rotation of the motion. Analogously to the above considerations, the following relationship then expresses the conservation of angular momentum: $m_{station} \cdot v_{station} \cdot r = m_{fuel} \cdot v_{fuel} \cdot r$. This allows r to be eliminated from both sides, resulting in the same formula as in the text. This consideration is valid because the mass of the space station is essentially concentrated at a distance r from the axis of rotation. If this were not the case, then the mass distribution in the space station would also have to be taken into account.

Holly: *"I'm coming in steep to get the last globe. I can't hold this course much longer. We'll break up at 200,000 feet."*
Bond *"A few seconds more."*
Bond shoots the laser and misses. The message "Malfunction" lights up.
Bond: *"Automatic firing system negative. Must be the heat. Switching to manual."*
Holly: *"Controls aren't responding. The wings are beginning to glow."* Bond: *Just hold her steady.*
Bond fires another shot with the laser and misses the third globe again.
Bond: *"Steady."*
Holly: *"It's entering Earth's atmosphere! James, this is our last chance."*
Bond: *"Steady."*

007 shoots again and hits at the last moment. All globes are destroyed and no longer pose a threat.

We would now like to analyse this dialogue in a little more detail. Holly Goodhead seems to be able to locate the globes by radar on a screen. This is of course possible and was already the standard in the military in 1979. Then we discover that there is a laser cannon on board which can destroy the globes. In principle, this would of course be possible, as we have already seen in our discussion of the Goldfinger laser. A laser with sufficient power, about 100 kilowatts, would melt the metal shell of one of the globes, and such a thing could be relatively easily integrated into a space shuttle nowadays. In 1979, this kind of device was not yet available, but already in 1983, the US launched the SDI program, whose goal was to develop ever stronger lasers to destroy approaching intercontinental missiles.[59]

However, there are three more oddities regarding Bond's laser cannon. Firstly, there is the fact that it is a blue laser. Although blue light is short-wave and therefore rich in energy, this does not mean that it is particularly well suited for melting metals. Long-wave light is more suitable for this because it is attenuated more strongly by the metal and therefore the energy is better absorbed. Blue light, on the other hand, would simply be reflected. So it would make much more sense to use red light for the laser cannons, just as Goldfinger does with his laser.

Secondly, it is very difficult to understand why the blue beams of the laser cannons are visible at all. With a laser pointer, only the point on the wall can

---

[59] In August 2014, the US Navy put the first laser weapon into operation. In published videos, the weapon is tested on unmanned aerial vehicles and inflatable boats, which start to burn after a short exposure time.

generally be seen, unless the air is dusty, for example, when a smoker blows some smoke into the laser beam or it passes through flying gold dust, as in Goldfinger's laboratory, in which case one can see the laser beam because the light is reflected by these particles. But it is not easy to explain why it is so dusty in space that we can see the blue light beam of the laser cannon so well. According to Holly, the spacecraft is at least 200,000 feet up. Since an English foot corresponds to 30.48 centimeters, this would be a height of 60 kilometers, after all. According to NASA, space begins at 80 to 100 kilometers,[60] but even at an altitude of 60 kilometers the particle density is so low that no appreciable light scattering can be expected. The only explanation for the blue rays is, therefore, that a satellite must have exploded at this altitude, its microscopically small pieces of debris making it possible to see the shots of the laser cannon so well.

Thirdly, the laser beam is highly divergent, i.e., it diverges relatively strongly when it leaves the cannon. This can be seen in Figure 4.12.

An analysis of the image shows that the laser beam fanned out two centimeters over a distance of about one meter. However, laser light is extremely concentrated and usually much more focused. It is precisely because of this that applications such as laser pointers are possible at all. So if Bond was aiming at a globe 500 meters away, the laser beam would fan out 10 meters over that distance. This would distribute the energy of the light over much too large an area and it would certainly not be able to penetrate the metal walls of the globes. So Figure 4.12 can't actually be showing the beam of the laser gun. Apparently, a normal divergent beam of light from a conventional lamp is emitted before the laser cannon is fired, as shown in Figure 4.12. This could be a safety device, like a kind of warning to announce that the destructive laser beam is coming right after it.

When it is chasing after the globes, the space glider heats up considerably. That would actually be the case. The Moonraker carrying Holly Goodhead and 007 enters the Earth's atmosphere at an altitude of about 100 kilometers and at a speed of about 30,000 km/h. Although the atmosphere at this altitude is still very thin, the space glider is already significantly slowed down by the surrounding atmosphere. The air in the shock front just ahead of the Moonraker or a Space Shuttle heats up considerably through compression. In addition, there is also the frictional heat, which can become very high at these speeds and also serves to brake the space glider. Its very high kinetic energy is

---

[60] The transition between the Earth's atmosphere and space is fluid. The Fédération Aéronautique Internationale defines the border with space at 100 kilometers above sea level. This is the so-called Kármán Line. At this altitude, the speed required to obtain lift for flying is as high as the orbital speed of a satellite, so that above this line one can no longer speak of aviation.

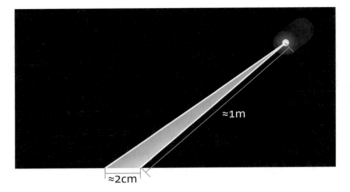

Figure 4.12 The laser beam exits the cannon of the space glider. Over a distance of one meter it fans out about 2 centimeters

thus transformed into heat, and it is braked without the use of fuel. Previous braking to lower speeds by means of rockets would require too much fuel and is therefore not carried out. A spacecraft thus heats up very strongly when it enters the atmosphere, so strongly that it would simply burn up without special protection. Therefore, all space capsules and orbiters returning to Earth have a heat shield. On the one hand, this heat shield should transfer the heat absorbed from the shock front as effectively as possible to the environment, and on the other hand, it should protect the spacecraft and its occupants from the heat by having very low thermal conductivity. This places enormous demands on the materials used for these heat shields, as they have to withstand temperatures up to several thousand degrees Celsius. Reusable heat shields such as the heat protection tiles of Space Shuttles usually consist of highly porous glass fibre materials bonded by sintering with a thin, dense, brittle, temperature-resistant borosilicate top layer.[61] Drax' Moonraker apparently also has such a heat shield. However, the fact that it still gets warm inside the space glider is quite realistic.

Holly also mentions that she had to choose a steeper re-entry angle into the atmosphere than usual to catch up with the globes. For the actual Space Shuttle and Apollo missions, the angle was usually between 6 and 7 degrees. If re-entry is too flat, the spacecraft will leave the atmosphere again. The astronauts could then attempt to enter the atmosphere again, but the target area would be so widely missed that it would no longer be possible to land in a controlled way. If the entrance is too steep, the thermal load on the space

---

[61] Particularly stressed parts of the Space Shuttle heat shield, such as the wing leading edge, were made of carbon fiber.

glider is too high—so Holly Goodhead is right in the remark she makes. But she can't have changed the angle of entry very much. When the Apollo space-craft re-entered after returning from the Moon, the angle of entry was ideally 6.5 degrees, with a tolerance of plus or minus 0.5 degrees. A greater variation in the angle of re-entry into the atmosphere would have involved far too great a risk. Half a degree of variation, however, would not have helped much in the pursuit of the globes, which seem to be following the shortest path to the Earth's surface.[62] All this suggests that the enchanting Ms. Goodhead was just trying to impress James Bond, because her initial skepticism towards 007 had meanwhile evaporated.

This is especially apparent at the end of the movie. Bond and Holly are in orbit in the Moonraker, while the ground base is happy about the happy out-come of the mission. We see Bond's Chief M, the technical whiz Q and the British Secretary of State for Defence Sir Frederick Gray along with numerous employees of the American CIA and the CIA director who want to set up a kind of video conference with the Moonraker.

CIA staff: "*Colonel Scott reports picking up two survivors—a tall man and a short blonde woman.*"

CIA Director: "*Right. OK. Dr. Goodhead reported their position. They're coming into range of our Pacific tracking ship. We'll have audiovisual within minutes from on-board monitors.*"

One recognizes the Moonraker flying through the picture, and the camera pans into the interior. There Bond and Dr. Goodhead can be seen lying naked in a kind of hammock, cuddling each other wildly.

CIA staff: "*Houston calling Dr. Goodhead. Houston calling. Confirm your position.*"

CIA Director: "*As this is the first joint venture between our two countries, I'm patching it to the White House and Buckingham Palace.*"

Gray: "*I'm sure Her Majesty will be fascinated!*"

CIA staff: "*We have audiovisual.*"

CIA Director: "*Ah, at last!*"

At first a noisy picture appears on the screen. Then one recognizes James Bond and Holly Goodhead closely entwined in a rather compromising position in the hammock. A murmur goes round the room.

M: "*007!*"

Gray: "*My God! What's Bond doing?*"

Q: "*I think he's attempting re-entry, sir …*"

---

[62] They must therefore have a super heat shield. Since they are much more compact than the space glider, this would be possible in principle.

5

Unbelievable Chases

Bartender: *"Yes, sir?"*
Bond: *"The lady'll have a Bacardi on the rocks."*
Anya Amasova: *"For the gentleman: Vodka Martini. Shaken, not stirred!"*
Bond: *"Touché!"*
(Quote from the 10th Bond film *The Spy Who Loved Me*)

How to Catch a Plane in the Air

After surviving a breakneck bungee jump from a dam at the beginning of the movie *GoldenEye*, James Bond breaks into the nearby chemical weapons factory in Archangelsk. There he meets his friend and colleague Alec Trevelyan, alias Agent 006, who is also in Her Majesty's service.[1] The two spies' mission is to blow up the entire facility. While attaching the explosive charges, however, they are discovered and attacked by Russian soldiers under the command of Colonel Arkady Ourumov. While Bond manages to place the charges and activate the timer, 006 is subdued by the attackers. Ourumov threatens to shoot him if James Bond does not surrender within ten seconds. Although 007 agrees, the colonel pulls the trigger shortly before the allotted time expires. Alec Trevelyan sinks to the ground, apparently fatally hit in the head.

James Bond manages to escape to a landing strip outside the facility. During the exchange of fire with the soldiers pursuing him, he sees a plane taking off,

---

[1] However, 006 turns out to be a double agent during the course of the movie and turns out to be Bond's opponent.

© The Author(s) 2020
M. Tolan, J. Stolze, *Shaken, Not Stirred!*, Science and Fiction,
https://doi.org/10.1007/978-3-030-40109-2_5

sprints towards it, reaches the plane, opens the side door, and jumps aboard. In the fight for control, James Bond and the pilot fall out of the starting plane. While the top agent is able to roll skillfully, the pilot collides with a soldier on a motorcycle. In a flash, 007 takes advantage of the situation, grabs the motorcycle and races after the driverless aircraft. The soldiers break off their pursuit in the face of this seemingly insane move—the runway finally ends at a deep abyss. In fact, a few moments later the pilotless plane falls off the cliff, followed two seconds later by James Bond on the motorcycle. In the air, 007 lets go of the motorcycle, puts out his arms and sails after the plane in the best superman manner. He gradually approaches and finally manages to reach his goal after 20 seconds. Despite an almost vertical dive, he manages to grab the still open side door and pull himself in.[2] Once in the cockpit 007 grabs the joystick and pulls it hastily toward himself. A few moments before the plane threatens to crash in the valley, Bond succeeds in pulling the plane up. As he flies over the factory, the explosives he has set there explode—Mission accomplished!

Wow—Bond was very lucky! But was it just luck, or was it the cool calculation of a brilliant secret agent? Would such a manoeuvre really be feasible in principle, or is it physically impossible?

We divide the scene into several different problems and discuss possible solutions. First Bond has to catch the plane in the air, then he has to get on, and finally he has to take control of the plane. Each of these three actions would appear to be extremely difficult. The first part of the maneuver is free fall. We neglect the fact that the engine of the airplane is still running as it falls and may generate some propulsion, and also the fact that the wings may provide some lift. All motions are thus considered to be dominated by free fall off the cliff. The other two effects could also be taken into account, but it turns out that they only make minor corrections. In contrast to the opening scene from the film *Moonraker, in* which James Bond, the pilot, and Jaws essentially fall vertically and thus perform a one-dimensional motion, 007 and the plane are now moving in two dimensions—horizontally and vertically. Physically speaking, the motions of James Bond and the plane are each the motions of objects thrown horizontally. The situation is shown in Figure 5.1.

In order to analyse these motions, we use the principle of independence, i.e., a two-dimensional motions can be composed of two independent one-dimensional motions in the horizontal and vertical directions. This principle of independence can be illustrated by a simple experiment. In Figure 5.2, one

---

[2] Even if the door wasn't open, James Bond could just open it in the air. But this takes a lot of time, and time is the only thing James Bond doesn't have in this scene!

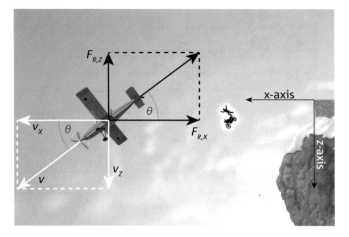

Figure 5.1 Directions of motion and forces when jumping off a cliff. *x* is the distance from the cliff and *z* is the distance fallen. The friction force F_R acts in the opposite direction to the falling motion. The horizontal and vertical components of the velocity *v* and air friction are denoted by v_x, $F_{R,x}$ and v_z, $F_{R,z}$. The angle θ between the direction of motion and the x-axis indicates the angle of inclination of the aircraft. A vertical fall thus corresponds to an angle of $\theta = 90°$

first sees two balls in a holder at the upper edge of the picture. If the two balls are released to fall to the ground at the same time, but the grey ball also receives a horizontal impulse at this instant, then it can be seen that they still fall equally fast. Although the light grey ball has travelled a greater distance, they are always at the same height at the same time. The two-dimensional motion of the light grey sphere is therefore nothing more than a vertical falling motion, with a simple, constant horizontal motion superimposed on it.

It follows from the principle of independence that Bond and the aircraft essentially fall vertically downwards. That they also move horizontally is clear, but the more important motion is the free fall. This is independent of the body itself, i.e., all bodies would fall equally fast if there was no air resistance. Thus without air resistance, James Bond could never reach the plane, because he would fall off the cliff two seconds later than the plane and always be those two seconds too late. Figure 5.2 illustrates this situation once again: just as neither of the two balls dropped at the same time reaches a greater depth of fall than the other, James Bond could not catch up with the plane's vertical drop in two seconds.[3]

---

[3] We assume that 007 has no external propulsion, such as some kind of jet propulsion on his back. With such an engine, he could of course easily catch up with the distance covered by the falling plane in two seconds.

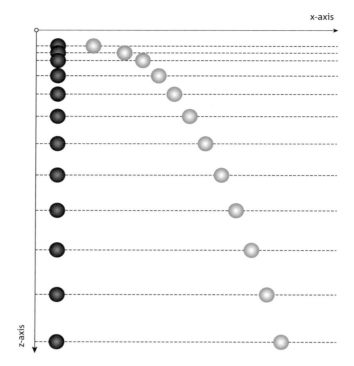

Figure 5.2 Principle of independence of motions. The dark ball falls perpendicularly to the ground, while the light ball has first received a horizontal impetus. It is clear that the two balls reach the same depth of fall z at the same time. The motion in the x-direction is simply superimposed on the falling motion

Taken the other way around, this means that Bond can only catch the plane if he uses the air resistance while falling, just like in the movie *Moonraker*. Since 007 and the airplane reach quite high speeds in free fall, their air resistance will be extremely important. Here, in particular, their streamlined shape, also known as wind slip, plays an important role alongside the other factors contributing to air resistance. From the drag coefficient, previously denoted by c_W, and the cross-sectional area in the direction of motion, we can calculate the effective drag on a body caused by the shape. As in the scene when Jaws could only catch up with James Bond in the air due to his greater weight, 007 now has quite a weight problem. The fact that heavy bodies fall to the ground faster than light ones is due to the fact that gravity must be compensated by air resistance. We have already analysed this in detail, in the discussion of the fall of Jaws in the film *Moonraker*. It ultimately leads to a higher rate of fall of the heavier body if the c_W values are similar. What James Bond lacks in weight compared with the airplane, he must gain through being more streamlined in order to take advantage of the air resistance and catch up with the airplane in the air. However, this is no walk in the park, as the aircraft is about 20 times heavier than the top agent. Bond would have to be 20 times more streamlined than the plane to fall as fast! This can definitely only be achieved with the help of British intelligence. For example, if Q has rubbed him with a special fluid that suppresses air vortex formation and thus greatly improves his streamlined shape. With Q, you never know! For us normal citizens this kind of escape is certainly not a solution, because even falling airplanes are very streamlined. The average citizen should therefore rather seek dialogue with his persecutors and in no case jump after an airplane!

But how exactly does James Bond take advantage of the air resistance to catch up with the plane? Shortly after leaving the cliff, the top agent pushes the motorcycle away. From this moment on he can consciously influence his air resistance and thus his trajectory in relation to the airplane. Bond's behavior is clearly visible in the film clip. First, 007 must reach as high a speed in the air as possible in order to catch up with the aircraft. He thus assumes a straight and narrow posture by stretching out his arms and legs as fully as he can. His air resistance is obviously much lower than that of the aircraft, because we can see that he is falling faster. Boarding, on the other hand, will only be possible if he manages to fly next to the aircraft at about the same speed. This requires braking in the air. Bond does this by moving into an upright position and thus greatly increasing his air resistance. He certainly doesn't have any other braking options. So the maneuver seems child's play for a top agent. But only detailed calculations of the flight paths of James Bond and the aircraft can check this more precisely and show whether boarding would actually be possible.

Figure 5.3 shows such a calculation of the flight curves. It can be seen that the resulting paths intersect. At the intersection James Bond and the plane are at the same distance from the cliff and at the same drop depth. This condition must, of course, be fulfilled before we can even begin. Calculating the intersecting flight curves is relatively easy. However, this is not all, because they must reach this intersection at the same time for any chance of success. One does not catch a train if one reaches the platform five minutes after its departure, even though one is then at the same place as the train was before.[4]

To get on board, James Bond and the plane must be at the same distance from the cliff and at the same drop depth at the same *time*. This calculation, however, proves to be extremely difficult. However, after some experimenta-

---

[4] There are completely different reasons why we can nevertheless often catch trains run by the Deutsche Bahn. This will not be discussed further here.

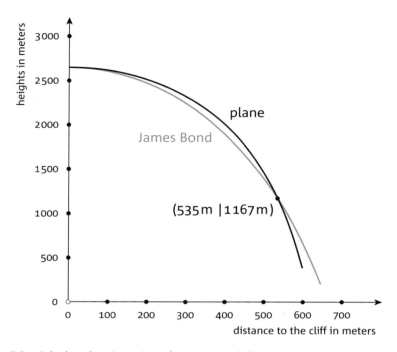

Figure 5.3 Calculated trajectories of James Bond (light curve) and the plane (dark curve) after jumping off the cliff. The two trajectories intersect, i.e., there is a point where 007 and the aircraft are at the same distance from the cliff and at the same drop depth. The graph shows that this intersection point is reached at a distance of 535 meters from the cliff and a drop depth of 1167 meters

tion with the values that James Bond can change in the course of his motorcycle ride and the flight through the air, a solution can be found.

At the beginning, when he is still on the motorcycle, he can change the initial speed with which he falls from the cliff, and adjust to the speed of the airplane if possible, because this speed will certainly be decisive for whether he can climb aboard later or not. In the air, the secret agent can change his streamline shape, and thus also his air resistance, in a steady manner, by altering his body posture. A well-trained top agent would certainly be able to control his body in such a way that he can intuitively correct his trajectory as needed. The result of an exact calculation, featured in Figure 5.3, shows a simultaneous intersection of the orbits under the assumption that the aircraft initially crashes off the cliff at a horizontal speed of 140 kilometers per hour.[5]

---

[5] This Pilatus Porter aircraft was provided for the shooting by the Swiss company "Air-Glaciers - Compagnie d'Aviation 1951 Sion". The plane is known for short take-off and landing distances and good reliability. It is therefore very popular among skydivers to help them out with their extreme activities.

Technically this initial speed is no problem for the aircraft, which is a Pilatus SC-6 or Pilatus Porter Turbo. Moreover, the motorcycle taken by James Bond, a Cagiva,[6] should be able to reach such speeds effortlessly. However, Bond must choose a slightly lower horizontal take-off speed than the aircraft, as his problem is not to catch the aircraft in the horizontal direction, but to make up for the two seconds of free fall in the vertical direction. If 007 were moving faster than the plane at the moment he left the cliff, he would pass hopelessly over the plane, since he must be much more streamlined. His problem is clearly not horizontal, but vertical. Nevertheless, he has to estimate the horizontal speed of the aircraft to within about 2 to 3 kilometers per hour in order to get anywhere near the aircraft.[7] James Bond could then correct a slight misjudgement of the speed during the flight by his posture, but this would certainly be rather nerve-racking. The secret agent must therefore remain extremely cool here!

007 could have pursued another strategy to catch the plane. Since his biggest problem is his low weight compared to the plane, he could have held the bike in the air to increase his total weight. However, this would increase the air resistance too much, because his streamlined shape would be significantly compromised. A greater total weight would allow James Bond to fall faster, but his streamlined shape plays a more important role here. Detailed calculations show that the influence of the c_W value far outweighs that of the greater weight. As Figure 5.4 shows, 007 has no chance of getting anywhere near the aircraft if he remains on the motorcycle, due to the much greater drag—not even at the maximum horizontal take-off speed of 180 kilometers per hour. Consequently, as was to be expected, our top agent's decision to push the bike off in the air as quickly as possible and adopt the most streamlined stretched-out shape is the only correct one. While he was sitting on the bike, James Bond must have calculated everything in his head with lightning speed in order to avoid making any decisive mistakes here. In no other scene is it so clear that the double-O agent must have studied a few semesters of physics, because the calculation of flight curves under the influence of air resistance is no mere high school subject.

So James Bond can actually catch up with the plane—who would have thought that? But is it also possible for 007 to get into it?

Contrary to the pictures of the film scene, in which it looks as if 007 could fly almost comfortably next to the plane, even with the most favourable cal-

---

[6] Presumably, it is a "Supercity 125" with a top speed of about 180 km/h.

[7] To estimate the speed of the airplane so accurately from the moving motorcycle would certainly be no problem for 007. Estimating speeds must be part of the basic training at MI6.

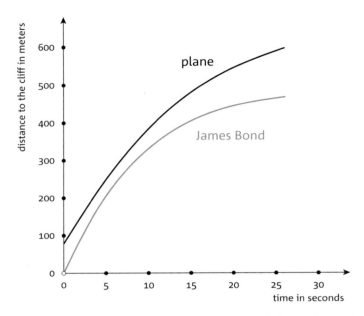

Figure 5.4 The distance of the top agent (grey curve) and the airplane (black curve) from the cliff as a function of time if Bond remains on the motorcycle in flight. We see that 007 has no hope of catching the plane because the two curves do not intersect. There is no time when James Bond and the plane are the same distance from the cliff

culation result, there is a very unhealthy relative speed of about 85 kilometers per hour at the intersection of the trajectories. Getting on board the airplane would be comparable with a crash at 85 kilometers per hour against a hard wall—007 would be in the airplane, but he would be dead! To prevent this, British intelligence must have invented a special suit containing implanted mini airbags. Q can easily be trusted to provide such a thing, because more often than not 007 has to survive just this kind of heavy impact. The high impact speed comes up with every calculation because James Bond is diving more or less beneath the flight path of the airplane and then has to enter laterally into it while it is falling almost vertically downwards; otherwise the two trajectories cannot be made to intersect. However, this configuration always requires a very high relative speed between the aircraft and the top agent.

Finally, we would like to clarify whether James Bond can still pull the plane up and save himself after a successful boarding. In the movie scene, 26 seconds elapse after the jump from the cliff until Bond finally grabs the control stick. He now has to pull the aircraft up out of its nosedive. After 26 seconds of free fall, its speed will have risen to an enormous 470 kilometers per hour. Aerobatic pilots can still manage such a manoeuvre at a speed of 300 to

400 kilometers per hour. Pulling the plane up is therefore the least problem of all the problems Bond has to solve until he reaches the cockpit. It goes without saying that he can fly any aircraft better than the best pilot of an aerobatic squadron.

26 seconds pass before Bond intercepts the plane. During this time he and the plane have already dropped more than 2200 meters in altitude, so this rescue takes up a lot of space. Does the cliff even offer enough space for this manoeuvre? In the vicinity of Arkhangelsk this is rather unlikely, because the town is a coastal town, hence only slightly above sea level. Accordingly, the makers of the film preferred to shoot on the Tellistock mountain in the Swiss Alps, which has a height of 2651 meters. This height was therefore used as the take-off height in the calculations. Whether the cliff there reaches down to sea level is questionable, but we may assume so since James Bond would appear to survive. So we see that there really would have been enough space for the spectacular dive. Of course, the scene wasn't really shot as shown in the movie. In fact, a stuntman on a motorcycle jumped after the plane. Instead of trying like 007 to catch it, however, the man used a hidden parachute after about 300 meters drop and preferred not to try to get into the machine by daring maneuvers in mid-air.

For all the calculations that can be seen in the graphs, a computer had to be used, because difficult differential equations had to be solved. 007 does all this in his head while he follows the plane on the motorcycle. Even experienced physics professors have to take their hat off to such an achievement.

Details for Know-It-Alls

In classical mechanics, the procedure for calculating trajectories is as follows. First, we determine all forces acting on the given object. These are added together to obtain the relevant equation of motion, which can then be solved by mathematical methods or with the help of a computer. Since this is a two-dimensional problem,[8] the directions of the forces must also be taken into account. Two forces are involved for the jump from the cliff. On the one hand there is the gravitational force of the Earth which always acts vertically downwards, i.e., the weight force, $F_G = m \cdot g$, where m is the mass of the falling body and $g = 9.81$ m/s^2 is the acceleration due to gravity. On the other hand, the

---

[8] Our world is, of course, three-dimensional. The jump from the cliff, however, can be described in terms of two spatial coordinates—the distance to the cliff and the height. If there were a strong side wind, the third spatial direction would also have to be taken into account. For the sake of simplicity, however, we assume total calm for our calculations.

falling body is affected by air resistance $F_R = \frac{1}{2} \cdot c_W \cdot A \cdot \rho_{air} \cdot v^2$. Again c_W *is* the resistance coefficient of the falling body, which depends exclusively on its shape, A is its effective cross-sectional area in the direction of motion, v *is* its instantaneous velocity, and ρ_{air} is the density of the air, which depends on the height at any given instant during the fall.[9] The quadratic dependence of v once again shows that the friction force increases very rapidly with increasing speed. The air resistance force is exerted in the opposite direction to the instantaneous direction of motion. Thus a part of the frictional force counteracts the horizontal motion, and a part counteracts the falling motion.

As shown in Figure 5.1, the trajectory can be described by two quantities: the distance from the cliff, which is referred to as x in the following, and the depth of fall, denoted by z. All quantities to be considered, such as speeds and forces, can be expressed relative to these two directions. Each velocity can thus be broken down into a horizontal velocity component in the x-direction, written as v_x, and a vertical component in the z-direction, describing the free fall, written as v_z. This procedure is always suitable for equations of motion, because a two-dimensional motion can be broken down into two one-dimensional motions. Mathematically, all this simply follows from the vectorial nature[10] of velocity and force, which ultimately expresses the principle of independence of motions in the x- and z-directions, as shown in Figure 5.2.

In the horizontal direction, i.e., in the x-direction, only the x-component of the resistance force acts. In the vertical direction, we have the gravitational attraction of the Earth acting downward, and the air resistance acting upwards, which is the z-direction. The x and z components of the forces are calculated using the angle of inclination Θ with some trigonometry, as can be shown in Figure 5.1. The equations of motion shown in Figure 5.5 can then be established and solved using a computer.

The solutions $v_x(t)$ and $v_z(t)$ are obtained for both Bond and the aircraft. Then the two functions $x(t)$ and $z(t)$ are obtained from those, likewise by integration over time. The latter indicate the distance from the cliff and the distance fallen as a function of time. For example, the functions $x(t)$ are shown in Figure 5.4 for Bond and the aircraft. If $z(t)$ *is* plotted against $x(t)$, the result is the trajectory $z(x)$ *as* shown in Figure 5.3. Now we calculate different functions $x(t)$ and $z(t)$ and trajectories and investigate whether there is a configuration in which James Bond and the plane are at the same distance from the cliff

---

[9] However, the density of the air was kept constant during the calculations for this scene. At an altitude of 2651 m, the air density changes noticeably, but this is not essential for the calculation.

[10] Vectors are physical quantities consisting not only of a number but also of a direction. Thus the velocity is a vector, because it has not only an amount, but also a direction. Time, on the other hand, can be completely described by a number. It is not therefore a vectorial quantity.

Equation of motion: $m\dfrac{d\vec{v}}{dt} = m\vec{g} - \beta v\vec{v}, \quad \vec{v}_0 = \begin{pmatrix} v_0 \\ 0 \end{pmatrix}$

In components, this equation is:

$\left.\begin{array}{l} \dfrac{dv_x}{dt} = -\dfrac{\beta}{m} v_x \sqrt{v_x^2 + v_z^2} \\[3mm] \dfrac{dv_z}{dt} = -g - \dfrac{\beta}{m} v_z \sqrt{v_x^2 + v_z^2} \end{array}\right\}$ **Coupled system of 1st order**
$\Rightarrow$ **nonlinear differential equations**
$\Rightarrow v_x(t)$ **and** $v_z(t)$

Figure 5.5 Equation of motion of a body of mass m falling under the influence of drag and with an initial horizontal velocity v_0. The quantity β is given by $\beta = \frac{1}{2} \cdot c_W \cdot A \cdot \rho_{air}$. The solutions of this system of differential equations are the velocities $v_x(t)$ and $v_z(t)$ of the body as a function of time in the horizontal and vertical directions

and at the same depth of fall after about 22 seconds. Figure 5.3 shows the results of such an investigation, and it is boarding time for the secret agent!

Since James Bond also influences his trajectory by changing the drag force, it was kept variable in the calculations of the flight curves in Figure 5.3. While the c_W *value* and the cross-sectional area in the direction of motion are almost constant for the aircraft, time-dependent values must be assumed for 007. This change is modelled by a mathematical function, which first provides a small c_W value and a small cross-sectional area in the direction of motion, making them both larger and larger during the course of the flight. This describes the deceleration of the top agent in the air when he stands upright just before boarding the aircraft. Figure 5.6 shows the product of the c_W value and the cross-sectional area in the direction of motion as a function of time.[11] Using two parameters, James Bond can have a considerable influence on his trajectory. He can first determine the difference between the maximum and minimum values of the product of c_W and the cross-sectional area. Furthermore, he can select the point in time at which his body "folds down", thereby specifically increasing air resistance. Both can thus be controlled in the calculation and precisely reproduced.

From the formula already given in previous sections for the terminal free fall velocity when there is air resistance, viz.,

$$v^2 = 2 \cdot m \cdot g / \left(c_W \cdot A \cdot \rho_{air} \right)$$

it follows that a heavy body falls just as quickly as a lighter body if the quantity $c_W \cdot A$ is correspondingly smaller. Since the Pilatus Porter weighs about 1500 kg, this quantity must be at least $1500 / 76 \approx 20$ times smaller for James

---

[11] For experts, the increase in air resistance was modelled using a hyperbolic tangent function.

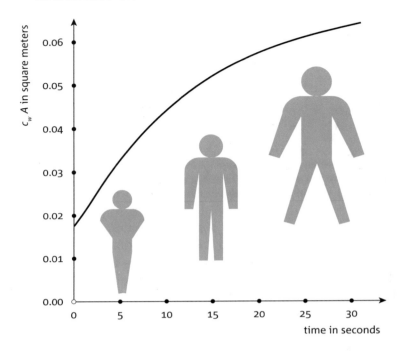

Figure 5.6 James Bond's air resistance depends on the product of the cross-sectional area A and his streamline shape, quantified by the c_w value. The curve in the drawing shows the time dependence of this variable, which is growing steadily as 007 slows down his flight to board the plane. The small pictures below the curve show how James Bond gradually straightens up and thus continuously increases his air resistance

Bond than for the falling airplane. Bond has to really stretch out to reduce his c_W *value* accordingly, especially as airplanes are usually already highly streamlined.

No Bridge: No Problem!

In the 1974 film *The Man with the Golden Gun,* James Bond performs one of the most spectacular and physically interesting car stunts ever shot. The scene begins with Bond's assistant Mary Goodnight trying to put a transmitter in the trunk of a car belonging to Scaramanga and his little assistant with the unusual name Nick Nack. Scaramanga pushes her into the trunk and locks her up. After she has informed 007 by means of a radio message from the trunk, he takes up the pursuit. But since Bond's car key is in Mary Goodnight's handbag, he has no choice but to "borrow" a new AMC Hornet Hatchback from a car dealer. On the passenger seat of the vehicle sits the well-known, slightly overweight Sheriff J.W. Pepper, who is already well-known from the film *Live and Let Die*. During the chase it happens that Bond and Pepper are on one side of a small river, Scaramanga and Nick Nack on the other side, with Mary Goodnight in the trunk. Since there is no working bridge nearby, there is a problem. 007 suddenly brakes and turns the car round with squealing tires. We then hear the following dialogue with Sheriff Pepper, who is looking somewhat intimidated on the passenger seat:

Pepper: *"Nearest bridge is two miles back."*

They pass by the remains of a collapsed bridge. It consists only of the two ends on opposite banks of the river, and these are slightly twisted and look like ramps. Bond makes an emergency stop, drives backwards, and comes to a halt in front of the ramp.

Pepper: *"Goddamn! What the... What's goin' on? What the hell are you doin' now, boy? The bridge is that way!"*

He sees the twisted ramp right in front of him.

Pepper: *"You're not thinking of...?"*
Bond: *"I sure am, boy! Ever heard of Evel Knievel?"*

Bond presses the accelerator pedal on his AMC Hornet and approaches the ramp at great speed. The car flies across the river. But since the ramps are twisted, the car turns once around its longitudinal axis during the flight, before landing on the other side of the river, about 16 meters away, as shown in Figure 5.7. Everything went well once again, and James Bond can continue the pursuit of Scaramanga, and finally save Mary Goodnight.

Here are some questions that need to be asked. Assuming that this "spiralling jump" really works, then it can only depend on the speed at which 007 drives onto the ramp. Everything else is fixed. If we know this speed—and we will get to know it soon—then everybody should be able to do this jump with their own car! Or is there another secret? Was this a trick? Perhaps, the scene was not a real stunt at all, just a piece of cinematographic deception.

A computer-generated scene, as we know from films from today, can be excluded immediately, because in 1974, these electronic brains were far from ready for such exploits. In fact, real leaps of this kind were demonstrated by various stuntmen at car shows in the Houston Astrodome in the USA in the

Figure 5.7 The flight of the AMC Hornet over the river. The car rotates around the longitudinal axis because the jump ramp is twisted. The take-off angle is $\alpha \approx 15°$, and the length of the ramp is $s \approx 3$ meters. The two ramps are located at a distance of $B = 15.86$ meters. What is required is the speed v_0 with which James Bond has to drive onto the ramp to reach the other side of the river

early 1970s. Raymond McHenry, an engineer and employee at Cornell University, New York, came up with the idea for this stunt and used a computer to calculate the shape of the ramps and the jump speed.[12] McHenry even patented this "spiral jump" and the corresponding twisted ramps in June 1974. How did James Bond know at what speed he had to drive onto the ramp to land safely on the other side of the river?

In a spiral jump, the center of gravity of the AMC Hornet travels through a simple parabolic trajectory,[13] but in addition to that the car rotates longitudinally around the center of gravity. By the way, this applies in general: any motion can always be described as a motion of the center of gravity with a superimposed rotation around this moving center of gravity. Rarely, however, can these two independent motions be recognized as beautifully as in the spiral jump that James Bond performs with Sheriff Pepper on board.

Since these two motions occur independently of each other, they can be examined individually, which considerably simplifies the description of the flight path. The jump over the river would have been possible without a rotation, of course, although not quite so spectacular. Physically, some additional conditions on the dimensions of the ramp can be deduced from the observed rotation, as we will see in a moment. The rotation has very little effect on the motion of the center of gravity, so the car essentially flies on a parabolic path,

---

[12] This computer was originally designed to simulate car accidents. Today, this task would be a trivial undertaking for the most mediocre of laptops. In principle, we don't even need a computer for these calculations …

[13] In a spiral jump, air resistance does not play a role. Firstly, the speed of the car is relatively low, which causes a low drag, and secondly, the motion does not last long enough, so the drag doesn't have long enough to inhibit the motion significantly.

just like a ball thrown through the air. With the ramps at an angle of 15 degrees and a distance of 15.86 meters apart, the car then jumps off at a speed of 63.5 kilometers per hour.

It remains to be clarified whether, at this optimum speed, the previously neglected rotational motion around the longitudinal axis of the car can actually be carried out, or whether there would be a rough landing on the roof. As already mentioned, due to the decay of the bridge, its ends were twisted so that the car is forced to rotate to the left when entering the ramp. The car then retains this rotation until it lands. The rotational motion is also assigned a speed, the so-called angular velocity, which is determined by the fact that the car has to rotate once around the longitudinal axis in the time from the jump to the landing.[14] This rotational speed is completely determined by the speed of the vehicle, the twist angle and the length of the ramp. The faster the car drives over the ramp, the greater its angular velocity. If the speed of the car is too low, the car does not turn far enough in the air and lands on the roof. If the speed is too high, the car turns too far and could hit the side. Thus the jump speed is not only responsible for the correct jump distance, but also for the correct rotation speed. The ramps provide a fixed angle around which the car must rotate. James Bond can precisely determine this angle by making a sharp assessment with his trained eye. The car has to rotate once around its longitudinal axis during the entire flight. Strictly speaking, Bond's eyesight provides a rotation angle of 360 degrees – 2 . 50 degrees = 260 degrees, since the two ramps are already rotated by 50 degrees. However, the car does not have to touch down perfectly with all four wheels at the same time. A small deviation from the ideal angle of 260 degrees can be tolerated in order to arrive undamaged on the other side of the river. Calculations show that James Bond and Sheriff Pepper should survive the jump on the ramp at an initial speed of 62 kilometers per hour without any great difficulties. The result of 62 kilometers per hour for the optimum initial speed to rotate around the longitudinal axis is well in line with the result of 63.5 kilometers per hour required to accomplish the purely parabolic flight. The two speeds are close enough together, indeed within all the tolerances available to Bond, and the jump works. If we consider that the car does not have to land completely horizontally, and also that it does not have to land precisely at a distance of 15.86 meters, then it turns out that the driver even has a tolerance of plus or minus 3 kilometers per hour. Our considerations so far would mean the following: Bond approaches the ramp at 63.5 plus or minus 3 kilometers per

---

[14] The angular velocity is the angle through which an object rotates per unit time.

hour and the car takes off. It turns once in the air around its longitudinal axis and then lands on its wheels on the opposite ramp at the other side of the river.

If that were all there is to it, then anyone could make that jump with their own car. But is this jump with superimposed rotation around the longitudinal axis really that easy? The answer is no! Because one important thing has not yet been explained: Why does the car actually rotate so evenly in the air and not go completely out of control, as things are usually shown in action movies where cars are thrown through the air? What influences the stability of a rotation?

Raymond McHenry said in his calculations that the car had to be perfectly balanced for this stunt. There is, of course, a physical reason for this. It is true that a body only rotates stably and evenly around an axis if its weight is distributed as symmetrically as possible around this axis of rotation. This is one of the reasons why a discus is so stable in the air during flight. A car, however, is not necessarily the ideal example of a rotationally symmetrical body. The roof and especially the occupants, who do not always have the same weight, are particularly problematic. These two factors ensure that during a spiral jump with a normal car, any small external disturbance, such as a little wind, would be sufficient to cause the car to rotate chaotically in the air—with the total loss of all involved. Therefore the roof of the AMC Hornet was replaced by a cardboard roof during the stunt. The car was divided lengthwise, and the tank, engine, and steering wheel were placed exactly in the middle of the car, while all seats and other "innards" were removed. Moreover, the stuntman had to be strapped in the middle of the car and drive it in a half lying position. The car had to be prepared in such a way that the weight distribution around the longitudinal axis of the car was as symmetrical as possible.

With such a carefully balanced AMC Hornet, it worked perfectly the first time, because the stuntman only had to keep as accurately as possible to the correct jump speed of 63.5 kilometers per hour and otherwise rely solely on the laws of physics. The latter will do the rest all by themselves.

Once again, the fact that James Bond has calculated all this lightning fast in his head and then managed to do it with an overweight passenger like Sheriff J.W. Pepper once again underlines his unrivalled class.[15]

---

[15] When jumping off the cliff after the crashing plane, James Bond apparently solves coupled non-linear differential equations in his head in no time at all. In comparison, the speed calculation for the spiral jump is a ridiculously simple undertaking.

Details for Know-It-Alls

Since no forces act on the car in the horizontal direction, it moves uniformly with the constant speed $v_{0,x} = v_0 \cdot \cos\alpha$. Here v_0 *is* the jump speed of the car and α is the angle of the ramp as shown in Figure 5.8. The horizontal distance $x(t)$ covered by the car in time t is thus given *by the* simple formula $x(t) = v_{0,x} \cdot t$.

The Earth's gravitational pull exerts a constant acceleration of $g = 9.81$ m/s^2 on the car in the vertical direction z. This causes the vertical speed of the car to change permanently according to the formula: $v_z(t) = v_{0,z} - g \cdot t$, where $v_{0,z} = v_0 \cdot \sin\alpha$ is the initial vertical speed component. Such a motion is called uniformly accelerated motion, since the speed always changes by the same amount per unit time. For the distance $z(t)$ travelled in the vertical direction, we then have $z(t) = v_{0,z} \cdot t - \frac{1}{2} \cdot g \cdot t^2$

If we solve the formula for $x(t)$ after the time t and insert this into the formula for $z(t)$, we obtain a quadratic function $z(x)$ for the trajectory followed by the car when it flies over the river:

Figure 5.8 The jump made by the AMC Hornet from the first ramp at angle α. The velocity v_0 *is* decomposed into a velocity $v_{0,x}$ in the horizontal direction and a velocity $v_{0,z}$ in the vertical direction

$$z(x) = x \cdot \tan\alpha - \tfrac{1}{2} \cdot g \cdot x^2 / \left(v_0 \cdot \cos\alpha \right)^2$$

In mathematics, a quadratic function is called a parabola. This special function describes the trajectory of any object thrown through the air when air resistance can be neglected.[16] The parabola can then be used to calculate the required jump velocity v_0 for a given jump width B and a given jump angle α. It turns out to be

$$v_0^2 = g \cdot B / \sin\left(2 \cdot \alpha \right)$$

From this we can easily calculate the jump speed to be $v_0 = 63.5$ km/h.

In addition to the translational movement, the twist on the ramp causes the car to rotate. The angular velocity is used for the physical description of a rotation. When driving on a ramp of length s, the car changes its angle of rotation φ in a time span $t = s / v_0$ determined by its speed v_0. The resulting angular velocity remains constant for the entire flight duration after leaving the ramp. However, it is also given by the quotient of the total rotation angle Φ during the flight duration T. Therefore, the condition $\varphi / t = \Phi / T$ follows immediately. The flight time T is directly linked to the jump distance and the horizontal jump speed $v_{0,x}$ of the car via $B = v_0 \cdot \cos\alpha \cdot T$. From the angles of rotation $\Phi = 260$ degrees and $\varphi = 50$ degrees, together with the length of the ramp $s = 3$ m, there is a second relationship between the jump velocity v_0 and the jump distance in addition to the equation for the parabola. There are therefore only very specific combinations of the jump distance and the jump speed which can lead to a successful spiral jump with given ramp parameters.

In the calculation, the translational motion of the car was considered completely independent of its rotational motion. This is also correct in principle, since only a little of the car's initial kinetic energy goes into rotation. The following energy balance always applies during the flight:

$$E_{kin} + E_{pot} + E_{rot} = \text{const.}$$

where $E_{kin} = \tfrac{1}{2} \cdot m \cdot v^2$ is the kinetic energy of the motion of the center of gravity of the car of mass m and speed v, $E_{pot} = m \cdot g \cdot z$ is the potential energy when the car is at height z above the ground, and $E_{rot} = \tfrac{1}{2} \cdot J \cdot \omega^2$ is the energy involved in the rotation of the car with moment of inertia J and angular veloc-

---

[16] The fact that a stone thrown through the air travels along a parabolic trajectory was already known to Galileo Galilei at the beginning of the 17th century.

ity ω about the longitudinal axis. The rotational energy was neglected in the previous consideration. If we take it into account, the calculation becomes much more complicated.[17] At the end of the day, a closer look reveals that the required take-off speed only increases by 7% from the 63.5 km/h previously stated, that is, to 67.5 km/h.

The fact that rotations are only stable if the weight distribution around the axis of rotation is as uniform as possible is not easy to understand. It is important to know that the angular momentum of a rotating body that is not subject to any forces remains constant. The angular momentum is a vector and thus has a direction in space. If this angular momentum is now parallel to the rotation axis, then the body rotates stably, otherwise the rotation axis must move around the angular momentum and the body will wobble. Now it can be shown that the angular momentum and the axis of rotation are parallel if the mass of a body is evenly distributed around the axis of rotation. But that requires higher mathematics, which of course our top agent masters perfectly.

A Car on Two Wheels

In the 1971 film *Diamonds Are Forever*, James Bond chases an old acquaintance, the megalomaniac Ernst Stavro Blofeld, who with his secret organization "Phantom" wants to build a laser weapon to blackmail the superpowers of the world. During his hunt, as so often happens, 007 comes into conflict with the strict interpretation of American laws, whereupon he is persecuted by various police forces. In a parking lot he shakes off his pursuers with some skilful driving manoeuvres. However, he then drives past the sheriff, who resumes the pursuit. James Bond now drives without realising into a dead end street, from which only a narrow footpath leads. But it doesn't matter—he uses a loading ramp on the left side of the road to bring the car into an inclined position and continues on two wheels. The width saved in this way allows him to pass through the narrow point without any problems, come out on the other side, tilt back onto his four wheels, and continue his journey. His followers try to do the same, but fail miserably and even overturn. How does it work? How can 007 get the car on two wheels. What exactly does he have to take into account? And for that matter, what did his pursuers do wrong?

A car standing on its side can only be balanced at a certain angle, namely, when the center of gravity of the car is above the support point on the road.

---

[17] But also not so complicated. This calculation belongs to standard at the TU Dortmund University in the exercises for the physics lectures for engineers in the first semester.

Only then can the gravitational force acting on the car not exert a torque that would return the car to a horizontal position. Therefore, once this angle is reached, the car could theoretically remain permanently in an inclined position. However, this condition is only as stable as an upright bicycle or a pencil standing on its tip, which is why we speak of an unstable equilibrium. The slightest deviation would immediately cause the car to overturn. Therefore it is not a solution for Bond to look for exactly this angle. It exists only in theory, and in a practical context, could never actually be achieved at full speed.

Since in reality gravity always produces a torque, another torque must counteract this in order to prevent the car from overturning. Of course, James Bond could try to balance the center of gravity with his own weight, but this would exceed even his abilities due to the almost twenty times greater weight of the car, a Ford Mustang Mach 1, weighing in at 1.4 tons. This undertaking would also be complicated by the presence of his lady passenger, who would certainly not be able to sit still for a driving style of this kind. James Bond uses the centrifugal force instead: in contrast to gravity, it acts laterally, as sketched in Figure 5.9. This means that its torque becomes maximum when the car approaches the equilibrium position. Sports cars are therefore lowered, i.e., built with a low center of gravity. They then offer the centrifugal force as little leverage as possible to prevent a rollover in tight bends. Since a centrifugal force only occurs on bends, James Bond can control it by steering. For example, if he steers to the left, as he does in the film and as shown in Figure 5.9, then the centrifugal force will push him to the right.

In addition, 007 must note that the centrifugal force depends not only on the strength of the steering, but also to a large extent on its speed. If he drives

Figure 5.9 A car in dynamic equilibrium. Gravity F_G and centrifugal force F_Z apply at the center of gravity S of the car and each generate a torque around the right-hand tires of the car. The angle α describes the position of the center of gravity relative to the supporting point and thus the oblique position of the car

too slowly, gravity wins and the car will fall back onto all four wheels. If he drives too fast, the centrifugal force will tip him over, which is what happens to his pursuers. However, James Bond is of course aware of this and therefore drives into the alley at just under 30 kilometers per hour. This is easily calculated by measuring the time in the film sequence and the length of the car. At this speed and given the low gradient of the ramp, the fact that the car also gets some additional momentum when starting to tilt plays a minor role.

So the secret of the stunt is that, at all times, the top agent skillfully balances a dynamic equilibrium between the leverage due to gravity and the leverage provided by the centrifugal force. This ensures that the car always remains on two wheels as long as it keeps moving. If it stops, the centrifugal force will go and 007 will no longer be able to balance anything at all, so the car will fall back onto its four wheels. That's how it is with cycling: only as long as we pedal and move forward do we remain in balance.[18]

How difficult is it to actually realize this stunt? If Bond misses the equilibrium position by several degrees, for example, he must take a correspondingly sharp curve with a small curve radius in order to generate a greater centrifugal force for compensation. Moreover, while driving, he will have to correct small deviations again and again, caused by smaller disturbances and his reaction time.[19] Figure 5.10 shows the relationship between the curve radius to be driven and the inclination of the car. It can be seen that, at a relatively large initial inclination of 75 degrees, a sharp turn with a radius of only 20 meters is required for the centrifugal force to keep the car in balance. Of course, a stuntman doesn't have to have Figure 5.10 in his head. He will react intuitively to small perturbations.

But what did the policeman in the car behind James Bond do wrong? Apparently, his centrifugal force was too great. The policeman tried to imitate James Bond's manoeuvre. After he has driven up the ramp at the same speed, he also steers intuitively to the left, as can be seen in the scene. However, he doesn't seem to be aware that the center of gravity of his car, including the occupants, is higher than that of the sports car driven by our top agent, so a much greater centrifugal force acts and the car eventually tips over. If the policeman hadn't steered so sharply to the left and left the ramp before reaching the end, he would have been able to follow James Bond safely and wouldn't have landed on the roof of his car.

---

[18] But riding a bicycle is still much more complicated, because there are more forces at work there.

[19] James Bond also has a reaction time, although it should be much shorter than the approximately 0.1 seconds of "normal" people.

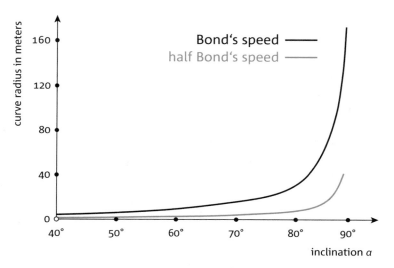

Figure 5.10 The inclination, i.e., the angle α from Figure 5.9, plotted against the curve radius to be followed to keep the car in balance. The upper curve is for the speed of James Bond's car in the movie *Diamonds are Forever*. The lower curve was calculated for half the speed. The greater the deviation from the unstable equilibrium position at 90 degrees and the lower the speed selected, the tighter the curves, i.e., smaller curve radii must be followed in order not to fall over

It remains to be clarified why the inclination of James Bond's car suddenly changes in the alleyway, because strangely enough the car drives into the alley on the right wheels and comes out again on the left wheels! How can James Bond do this? The obvious explanation would of course be that the scene was shot several times and the producers unfortunately connected two film segments with different lateral positions. In order to cover this up, an intermediate scene was shot afterwards, in which you can clearly see how the inclination of the car changes in the lane.

However, a completely logical explanation for this would also be that there are two ramps arranged in the lane as shown in Figure 5.11. The lane would then have to be wider again after the entrance, but this can neither be confirmed nor excluded from what we see in the film. James Bond drives into the lane on two wheels, then shifts the weight in the middle of the lane and tips the car over. Then he drives to the other ramp on the other side of the lane and tilts the car in the other direction, using all the effects already mentioned. And voilà! Bond comes out of the alley on the other two wheels, and everything has a perfectly logical explanation—without reference to any film errors!

So the car scene could have been shot just like that. Driving a car in an inclined position is a common stunt that can be seen in many movies. It is not

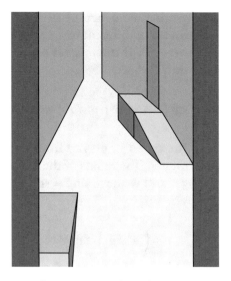

Figure 5.11 In the narrow alleyway, a number of ramps may have been positioned at a wider point. After entering the lane, 007 can use the left ramp to raise the two left wheels of the car, then the right ramp to raise the two right wheels before exiting

considered particularly difficult among stuntmen.[20] James Bond even masters this manoeuvre so perfectly that in the film *License to Kill*, he similarly tilts a large tanker truck loaded with explosive fuel to avoid a Stinger missile pointing at him.

Details for Know-It-Alls

If forces are applied to a body that can rotate around a point, they generate a torque. In general, a body is in equilibrium when all acting forces and all acting torques balance each other. The torque is defined as the product of the acting force and the distance of the point of action of the force from the pivot point. For our car, however, this distance also depends on the angle between the direction of gravity, acting vertically, and the lever arm from the fulcrum to the center of gravity of the car. In Figure 5.9, the corresponding angle is $\beta = 90° - \alpha$. In this case, the pivot point is the support point of the tire touching the ground. In the case of the unstable equilibrium position, this angle is equal to $0°$, i.e., $\alpha = 90°$. This corresponds to the minimum torque that can be exerted by gravity F_G, i.e., zero. The maximum torque results for $\alpha = 0°$. The

---

[20] The stunt, however, also requires a car with a differential lock. This ensures that the car is driven even when there are only two wheels on the ground.

centrifugal force F_Z, on the other hand, acts at an angle of 90° to gravity and thus has a different lever arm and a different torque. This torque is highest for $\alpha = 90°$ and disappears for $\alpha = 0°$. If the torques are now equated, it follows from the geometry in Figure 5.9 that

$$m \cdot g \cdot \sin \beta \cdot a = m \cdot \left(v^2 / r \right) \cdot \sin \alpha \cdot a$$

where *a is* the distance of the center of gravity from the contact point of the tire around which the car rotates. The radius of the curve that the driver travels at speed *v is* denoted by *r,* and we have $\sin\beta = \sin(90° − \alpha) = \cos\alpha$ whence the curve radius is given by

$$r = \left(v^2 / g \right) \cdot \tan \alpha$$

This relationship between the curve radius and the inclination is shown in Figure 5.10.

Now we explain how we determined the speed *v* of the car. We measure the time *t* between the front end of the car reaching a certain point and the rear end of the car passing this same point. The length of the Ford Mustang is known to be 4.8 meters. The time *t* could be measured with a stopwatch in front of the television. But that would be far too imprecise. It is better to decompose the scene into its individual images and to extract the relevant time from them. A film is composed of 25 frames per second. The passage of the car takes exactly 16 pictures. So the car needs a time of 16/25 s to drive through the 4.8 m long stretch. This results in the speed *v* of the car being v = 4.8 m / (16/25) s = 7.5 m/s = 27 km/h.

When it leaves the ramp, the car has already reached an inclination of about 45°. In order to get into the equilibrium position without having to steer too hard, James Bond should reach at least 50°. To add this missing 5°, he uses the momentum gained while going up the ramp. The car is rotated when the ramp is raised.[21] The resulting rotational energy can be derived from the mass distribution of the car, including its occupants, and the angular velocity of the rotation. It is determined once again by a simple image analysis of the car on the ramp. The stored energy is used to raise the car's center of gravity when it is lifted by moving up the ramp. If everything is now calculated, a height change of the center of gravity of about 5 cm results. This cor-

---

[21] Reference should also be made here to the statements for experts in the section entitled "Rollovers and low blows".

responds exactly to the desired change in the inclination from 45° to 50°. Once again, it's amazing how quickly and precisely James Bond calculated and implemented his actions!

Running on Water!

In the film *Live and Let Die* Bond is abandoned by the villain Dr. Kananga on a crocodile farm in a Bayou.[22] He lands on a tiny island that is disconnected from the land. Many hungry crocodiles are swimming in the water around it, slowly coming ashore and pushing him further and further back. Then Bond comes up with the idea of attracting a nearby boat with his magnetic watch. We will take a closer look at this watch later. This seems to work at first, but then the plan fails because the boat is moored. The crocodiles are getting closer and closer. 007 can escape the predicament by a courageous run over the backs of some crocodiles and thus reach the safety. It almost looks like our top agent can walk on water. Can he really do that? Can James Bond get ashore by walking on the bodies of crocodiles without them sinking into the water first? In principle, could he actually walk on water?

For this we have to consider which force could balance 007's weight to prevent him from sinking in. There are two effects here. First, a certain amount of water is displaced per unit time by his feet. This water is accelerated and so a force is exerted. However, it turns out that this effect is quite small. The friction force is much stronger again, because the foot has to move through the water when he is running. The force caused by the friction could in principle balance the weight force and thus ensure that James Bond could walk on water.

However, in practice this would already look somewhat different. Using realistic numbers, i.e., 76 kilograms for Bond's weight, a foot area of about 100 square centimeters, and a c_W value of the feet of about 0.8, the result is that 007 would have to race over the water at a speed of 70 kilometers per hour in order not to sink. But even James Bond can't do that. If, on the other hand, he increases his foot area tenfold by stepping on the crocodiles, this value is reduced to a realistic 22 kilometers per hour. James Bond could achieve such a speed very easily. And indeed, this scene was shot with a stuntman walking over trained crocodiles lying in a row. The problem was not so much the sinking in of the crocodiles or the speed with which the stuntman had to run, but the crocodiles were not as calm in the water as they were expected to

---

[22] Bayou is a common term in the southern states of the US for stagnant or slowly flowing water. Bayous were often the only traffic routes, especially in the difficult to access marshlands of the Mississippi delta.

be. In fact, this scene had to be rehearsed more than five times by the crocodile farm owner and stuntman Ross Kananga before it was finally captured.

By the way, there is actually an animal that can walk over water. This is the Jesus Christ lizard from the basilisk family, which inhabits the tropical rainforests of Latin America. Due to its low mass of about 100 grams, a speed of only 9 kilometers per hour suffices for it to achieve this feat. However, for such a small lizard this is already a very considerable speed, so it uses this ability, just like James Bond, only in those rare cases when it is in great danger.

Details for Know-It-Alls

Let us now take a closer look at which effects are important when running on water. If you try this, the weight of your body will just lead to your sinking in. So there must be forces that balance the weight of the runner. On the one hand, there is the recoil generated by the water displaced downwards by the feet, as in the case of a rocket. It's pretty small, though, as we can easily calculate. Much greater is the friction force, which results from the fact that the foot must force its way through the water. This friction force has already been discussed several times:

$$F_R = \tfrac{1}{2} \cdot c_W \cdot \rho \cdot A \cdot v^2$$

It depends on the cross-sectional area A *of* the feet, the density ρ of the liquid on which the person walks, the c_W value, and the square of the speed v^2 of the person's foot.

As always, the weight force is

$$F_G = m \cdot g$$

where m is the mass of 007 and g the acceleration due to gravity.

If we now equate these two forces, i.e., $F_R = F_G$, then this equation can be solved for the speed. We thus obtain the minimum value that a runner would have to achieve in order not to sink into the water. However, we must still consider the geometry of the foot immersion, which is relatively complicated.[23] We find that the friction force is only half as great, i.e., $F_R = \tfrac{1}{4} \cdot c_W \cdot \rho \cdot A \cdot v^2$

---

[23] More detailed information can be found in the original publication on the Jesus Christ Lizard running over water by J.W. Glasheen and T.A. McMahon, *A hydrodynamic model of locomotion in the Basilisk Lizard*, Nature Vol. 380, p. 340–342 (1996).

From this the square of the required speed can be calculated:

$$v^2 = 4 \cdot m \cdot g / (c_w \cdot \rho \cdot A)$$

In this formula m = 76 kg was used as Bond's mass, c_w = 0.8 for his foot forcing its way through water, and ρ = 1000 kg/m^3 as the density of water. An area A = 100 cm^2 was assumed for the foot, which leads to v = 70 km/h, but with the increased value of A = 1000 cm^2 when Bond runs over the crocodiles, the result is an attainable 22 km/h.

6

Gimmicks and Gadgets

Bond: *"I want to see Valentin Zukovsky."*
　Gangster: *"Impossible!"*
　Bond: *"Vodka martini. Shaken, not stirred. Tell him James Bond is here—now!"*
　(Quote from the 19th Bond film *The World Is Not Enough*)

Every James Bond fan has noticed that our elegant secret agent wears very interesting watches and glasses. He doesn't always put so much emphasis on style as on functionality, since the watches and glasses usually contain a technical "gimmick" that can save 007 from many a hopeless situation. But we do not want to discuss the usual equipment of a secret agent with watches and ballpoint pens containing miniature microphones and cameras. A schoolkid wearing the "ultra-modern" quartz watch with liquid crystal display, whose built-in camera is praised in the 1983 film *Octopussy* as a technical masterpiece, would certainly not find a girlfriend in the schoolyard today. It is also easy with today's technical possibilities to integrate a Geiger counter into a watch, as so proudly presented by Q in the film *Thunderball*. Here, one simply uses semiconductor detectors in which the counter tube consists of a germanium crystal where charge separation by the incident ionizing radiation takes place. One such watch was launched on the market by a Swiss company in 2003 for 1100 US dollars. This watch should also be water-resistant at depths of up to 100 meters: *"It's waterproof, of course"*, as Q remarks when presenting the watch (Figure 6.1).

© The Author(s) 2020
M. Tolan, J. Stolze, *Shaken, Not Stirred!*, Science and Fiction,
https://doi.org/10.1007/978-3-030-40109-2_6

Figure 6.1   James Bond (Pierce Brosnan) with the ingenious super tinkerer Q (Desmond Llewelyn), who always supplies 007 with his little gadgets. Scene from the movie *Tomorrow Never Dies*

We would like to present what would appear at first glance to be two even more ordinary watches, which only reveal their true secrets on closer inspection. They come from the movies *Live and Let Die* from 1973, the first movie in which Roger Moore played the top agent, and *The World is Not Enough* from 1999 with Pierce Brosnan in the leading role. In the latter, 007 also sports "X-ray glasses", which cry out for a more precise analysis. The sunglasses from the 1985 film *A View to a Kill* are also very impressive on closer inspection. And last but not least, we shall investigate what is probably the most spectacular gimmick Bond has to offer in these movies: the rocket backpack presented in 1965 in the film *Thunderball*, with which he escapes from danger in the opening sequence.

The Distance Makes It Possible: A Magnetic Watch

In the film *Live and Let Die* James Bond receives instructions from his superior M for a new mission. This time the villain Kananga is to be convicted. M drinks a cup of tea during the conversation, and Miss Moneypenny hands over some material for the upcoming mission to the agent standing in front of her in a yellow bathrobe:

Moneypenny: *"I've got your ticket to New York. Q has repaired your wristwatch. And some background on San Monique."*
Bond: *"Thank you, Moneypenny."*
M: *"I'm sure the overburdened British taxpayer will be fascinated to know how the Special Ordnance Section disburses its funds. In future, Commander, may I suggest a perfectly adequate watchmaker just down the street."*
Meanwhile, Bond pulls out a button on the watch, which is the luxury model of the luxury brand Rolex. About three seconds later, the spoon that M is using to stir his tea is attracted by the watch. It flies about a meter through the air and sticks to Bond's watch.
M: *"Good God!"*
Bond: *"Pulling out this button turns the watch into a hyper-intensified magnetic field, powerful enough to deflect the path of a bullet—at long range or so Q claims."*
M: *"I feel tempted to test that theory now. If you don't mind, Commander, my spoon."*
Bond: *"Sorry, sir."*

Q obviously didn't just repair the hands or insert a new battery, he turned the watch into a super magnet. But could this magnetic watch really work like this[1]?

The watch has been converted into an electromagnet—that much is immediately clear. Electromagnets are nowadays everyday objects. In physics classes, everyone must have coiled a wire around a nail and connected it to a battery. This is already a primitive electromagnet that could well attract a spoon, as we will see. In an electromagnet, a magnetic field is generated by an electric current. This also explains why, unlike a permanent magnet such as a piece of magnetized iron, it can be switched on and off. The magnetic watch is therefore an electromagnet, which could then naturally attract a spoon made of iron or steel.

"It works!" one might say of this gadget, if the principle can be explained *qualitatively*. But physics is a science that delivers *quantitative* results. This means that numbers are always calculated in a physical analysis. And we can also easily formulate this quantitatively.

For the spoon to be attracted to the watch, the magnetic force of attraction must be greater than the force of gravity holding the spoon on the saucer. These are the two forces acting on the spoon, as shown in Figure 6.2. Gravity is easy to calculate. With a spoon of mass of ten grams, this results in a value

---

[1] The magnetic watch was voted the most popular James Bond gadget by fans in 2002.

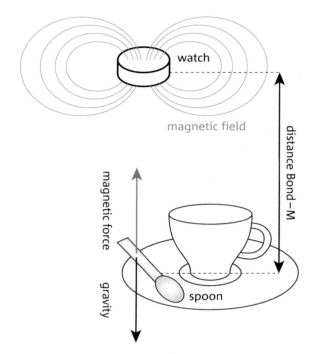

Figure 6.2 Illustration of the forces acting on M's spoon. If the magnetic force of attraction is greater than gravity, the spoon will be attracted by the magnetic watch

of 0.1 newtons. That's not much. To determine the magnetic attraction, the distance between James Bond and the spoon must be estimated. If one analyses an image of the film scene, two thirds of the top agent can be seen. Since James Bond is 1.83 meters tall, this distance is about 1.2 meters. The distance between 007 and M is the same, also 1.2 meters. Bond's Rolex watch has a diameter of about two centimeters. This information is also required for the analysis.

Now we have everything we require to formulate the question quantitatively and analyze the scene in detail. Since it is an electromagnet in the watch, we now ask more precisely: How large would the current that flows through the watch have to be in order to generate a magnetic field that can attract an iron spoon from a distance of 1.2 meters? We assume that M is stirring his tea with a cheap iron or steel spoon and not with an expensive silver spoon. Since he also has the British taxpayer in mind, this assumption seems reasonable. Bond could never attract a silver spoon with his watch—that's clear anyway. So how much electric current does the battery in his watch have to provide?

Our first consideration is that James Bond's watch is obviously a very expensive Rolex, as just the sight of it makes M think directly of the British taxpayer. Therefore, the function of the magnetic watch will first be explained with a structure in which it does not need to be opened or modified in any way. We assume that Q simply hid a copper wire coil under the watch—nothing more. Now it is possible to calculate the current that would have to flow through this watch to attract a 10 gram teaspoon at a distance of 1.2 meters. This results in a value of about 4.5 billion amperes! This is a quite remarkable value! Let us illustrate with a few comparisons. A torch is operated with about 0.2 amperes and an electric train locomotive with about 300 amperes, while a current of 100,000 to 1,000,000 amperes flows for a short time in a flash of lightning during a thunderstorm. Electrical currents of several billion amperes are really huge and do not even appear in the most extreme situations of nature. A battery that can supply even the slightest fraction of such a current is therefore completely unthinkable—even for Q, who has already made the impossible possible.

There would also be another great problem. An electric current generates not only a magnetic field, but also heat. The electric current flowing through a wire inevitably heats it up. Many electronic devices therefore have a fan or other type of cooling so that this heat can be dissipated into the environment. A calculation shows that the watch would heat up to a temperature of several trillion degrees Celsius[2] if 4.5 billion amperes of electricity were to flow through the wire! As a consequence, James Bond would be disassembled into his atomic components shortly after turning on the watch and evaporate instantly. So that's certainly no way to do it! But there is still much room for improvement, as we shall now see.

As a second attempt we will now unscrew the magnificent watch and remove all its innards. In an electromagnet, wire is wound into a coil around an iron core as this massively amplifies the magnetic field.[3] The shape of the coil is determined by the watch, because the diameter and height of the coil must of course fit into the watch case. The latter has a diameter of two centimeters and a thickness of at most one centimeter, as shown schematically in Figure 6.3. The strength of the magnetic field due to the coil can only be influenced by the number of turns, the electric current flowing in it, and the

---

[2] In this case, it does not really matter whether the temperature is given in degrees Celsius or kelvin.

[3] Iron is a ferromagnetic material and is effectively made up of microscopically small magnets. All these small elementary magnets are aligned by an external magnetic field. This drastically increases the external field. Only a few materials have this property. However, when all elementary magnets are aligned, the material is in magnetic saturation. An increase in the current would not cause any further noticeable amplification of the magnetic field.

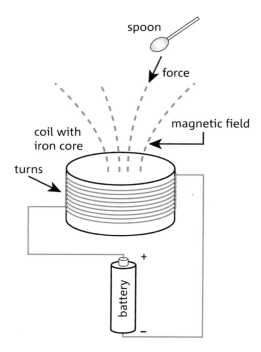

Figure 6.3 Schematic representation of a cylindrical coil to be integrated into the magnetic watch. The magnetic field is generated by a current supplied by a battery and exerts an attractive force on the spoon. Note, that the magnetic field lines are not complete. They are always closed and must also run through the coil

material enclosed in the coil, usually iron. We now assume that a coil with 100 windings of thin copper wire could be used. Since we assume a watch about one centimeter high, this number is already rather optimistic. This results in a magnetic field that is 100 times greater than the one produced by the watch previously, and thus in a current that is 100 times smaller than the current needed to attract the spoon. 45 million amperes are still unrealistically large. Furthermore, as shown in Figure 6.3, the copper windings can be wound on an iron core. This increases the magnetic field again by a factor of 5000 and the current required to attract the spoon is reduced to "only" 9000 amperes. If Q could actually develop a battery that could supply this current, then the watch would work.[4]

Nevertheless, 9000 ampere is still such a huge current that, apart from the fact that Q must have invented a super battery, the heat developed in it would

---

[4] And he would certainly receive the Nobel Prize, because new physical laws would have to be discovered to achieve that!

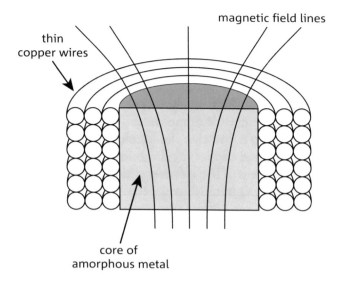

thin
copper wires

magnetic field lines

core of
amorphous metal

Figure 6.4 Cross-section of the improved coil. Calculations show that three-layer windings with 600 turns of wafer-thin copper wires give realistic values for the magnetic watch

still be problematic. The 9000 amps of the special battery would heat the watch to a temperature of about 40 million degrees Celsius.[5] As a consequence James Bond would still be disassembled shortly after switching on the watch and evaporate once again. So it couldn't work that way either.

The magnetic watch must therefore be improved again. First, we replace the iron core with a material which, when wrapped in copper wire, even more efficiently amplifies magnetic fields. Here amorphous metals are suitable. These are used to make high performance magnets. They provide a field enhancement factor of about 500,000 instead of 5000 for iron. Furthermore, we assume the diameter of the watch to be three centimeters instead of the more realistic two centimeters. However, the thickness is reduced to half a centimeter. The number of turns must also be increased considerably. Instead of single-layer winding, a multi-wound coil is used, as shown in Figure 6.4. Now three layers of 600 turns of extremely thin copper wire are used.[6] This means a total of 1800 turns, instead of the previous 100.

---

[5] For comparison, this is roughly the temperature that prevails inside the sun.

[6] The thickness of the wire would in fact only be 0.01 millimeters if 600 turns were to be wound in three layers onto a core of a watch that was 0.5 centimeter thick. However, it would still require a lot of research to produce such thin wires that do not melt at the high currents necessary here.

These two improvements result in a gain by a factor of $100 \cdot 18 = 1800$. This means that a current of only about five amperes would now be needed to attract the spoon from a distance of 1.2 meters, while the temperature of the watch will rise to 250 degrees Celsius. Perfect! 250 degrees Celsius would be unpleasantly hot on the wrist, but a top agent like James Bond will be able to take that for a short time. Alternatively, this temperature could be shielded by a ceramic plate on the underside of the watch.

Nevertheless, we must admit that the watch would certainly be very difficult to make. The required copper wires would be extremely thin and would have to withstand a current of five amperes without melting. This seems completely unrealistic unless Q has invented a special copper alloy that can withstand it. A current of five amps is also very large, considering that it is supposed to be generated by a tiny battery in the watch. Here too, Q would still have to fiddle around with a lot of details in order to achieve this. But there is a deeper reason why this magnetic watch, which at first sight seems so common, will never exist in reality.

To answer this, we need to study the strength of magnetic attraction forces more closely. First of all, this depends on the diameter of the coil producing them. This dependence is important because it goes as the fourth power. This means that a watch with twice the diameter does not just generate twice the force of attraction, but $2^4 = 16$ times. Increasing the diameter of the watch from two centimeters to three centimeters not only ensures that more copper wire will fit into it, but also increases the attractive force by a factor of $1.5^4 \approx 5$ due to this effect.[7]

The influence of the distance between the spoon and the magnetic watch is even greater. Magnetic attraction forces quickly drop to tiny values with increasing distance. Strictly speaking, a magnetic attraction force at ten times the distance will have decreased, not to one tenth of its original value, but already to one ten millionth. The force decreases with the seventh power of the distance between the magnet and the object to be attracted. A comparison shows the importance of this dependency even better. A small electromagnet, of the kind one might make with some copper wire and an iron nail, can easily attract an object of ten grams, for example a teaspoon, at a distance of one centimeter. This corresponds to our everyday experience. At a distance of ten centimeters, on the other hand, the same electromagnet can only attract an

---

[7] The fact that the diameter plays a major role can also be seen from the fact that a spoon can actually be attracted from a distance of one meter by a large magnet, of the kind used in junkyards. This experiment was carried out by the authors at a Dortmund junkyard in 2007. Such a magnet has about 50 times the diameter of the James Bond watch magnet. But the magnetic attraction at a distance of one meter is $50^4 = 6,250,000$ times greater!

object of 10/10,000,000 = one millionth of a gram, or one microgram, which corresponds approximately to the mass of a grain of sugar. At a distance of one meter, the force would be again ten million times smaller, and then practically nothing could be attracted. Even at a distance of only one meter, there is almost nothing left to feel of the force emanating from the magnet. This shows that Bond's magnetic watch has to generate an extremely strong magnetic field in order to attract a spoon at this distance. This is therefore the real reason why such large currents have to flow through the watch. These currents ensure that at least a little magnetic field arrives at a distance of one meter. One meter is a huge distance for a magnetic field. So this is the central problem in the realization of Bond's magnetic watch.

But now we have to investigate why Q actually developed this device. Bond says it's possible to deflect bullets. Can that really be true? First of all, there is a fundamental problem here. James Bond can only attract objects with the magnetic watch, not repel them, unless the bullets are themselves magnetic. However, this characteristic is rather unlikely for a pistol bullet. So when James Bond aims his watch at a bullet, he will be guiding it directly toward himself! That would be good news for his enemies, but less good for him.

But then why would it be useful to deflect bullets? Perhaps Q meant that the watch could be used to protect a third person from being shot. Perhaps he intends it for a situation like the one in Figure 6.5.

A villain aims straight at his victim, and James Bond stands perpendicular to the line of fire. Now when Bond turns on his watch so that the bullet passes through the magnetic field on its way to the victim, it will indeed be slightly deflected.[8] In the magnetic field its flight direction will change by a small angle. It will then fly on at a constant speed in the new direction. If the magnetic field of the watch is strong enough, the bullet will ideally fly past the victim. The deflection angle depends on the strength of the magnetic field and of course on the speed of the bullet. The distance between the victim and the villain is also essential for working out whether the victim is ultimately missed. As Figure 6.6 illustrates, the longer the bullet continues to fly after being deflected by the magnetic field, the further it moves away from the original line of fire and the more likely it is to miss the victim.

Figure 6.7 shows the result of calculations for the minimum distance between the victim and the villain to ensure that the bullet just misses its target when James Bond is standing at a certain perpendicular distance from the line of fire with his magnetic watch. A watch with the characteristics shown in

---

[8] Bullets are often made of lead, but usually have an iron core. Pure lead bullets could only be deflected imperceptibly by the magnetic watch.

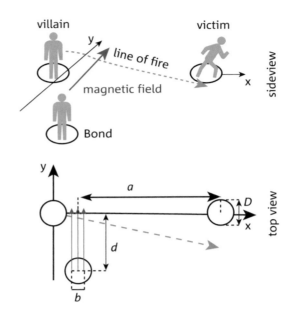

Figure 6.5 James Bond uses his magnetic watch to protect a person the villain is trying to shoot. He is at a distance *d* from the line of fire, and the villain is at a distance *a* from the victim, as can be seen better in the top view (lower figure). Bond deflects the projectile with his magnetic watch in such a way that it misses the victim

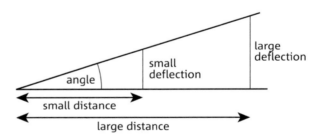

Figure 6.6 The same deflection angle produces a small lateral deflection at a small distance, but a correspondingly larger deflection at a large distance

Figure 6.8 is assumed. The distance between 007 and the path of the shot is read off on the horizontal axis, the necessary minimum distance between the victim and the villain on the vertical axis. The diagram shows, for example, that if the top agent with his magnetic watch is standing one meter away from the path of the shot, the victim must be at least ten meters away from the bad guy for the bullet to miss. That sounds pretty good. However, the necessary distance quickly increases as James Bond moves away from the line of fire. If he stands about 1.2 meters away, the distance between victim and shooter

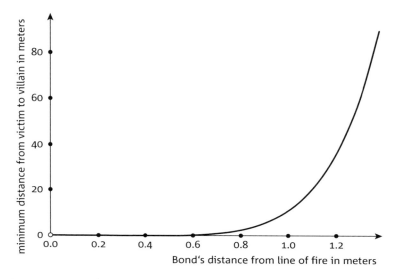

Figure 6.7 James Bond's magnetic watch is positioned at a certain distance from the villain's path of fire (horizontal axis). The dark line then describes the minimum distance (vertical axis) between the victim and the shooter in Figure 6.5 so that the villain's bullet misses its target. For example, you can read from the graphic that if Bond is 1.2 meters away from the line of fire with his watch, the victim must be at least 30 meters away from the villain so that he can still be rescued by 007 with his magnetic watch. One recognizes that this minimum distance increases very strongly with increasing distance of the secret agent to the firing path

diameter	3 centimetres
thickness	0.5 centimetre
coils in three layers	1800, copper wire
material of the core	metallic glass Factor 500 000
delay time (90 % of maximum strength)	3.5 seconds
operating current	5 amperes
operating temperature	approx. 250 degrees celsius

Figure 6.8 Summary of the most important technical data concerning a magnetic watch that would work just like the one in the film *Live and Let Die*

must already be 30 meters. If the distance between 007 and the shooting path is just 1.5 meters, the victim would have to be at a distance of more than 80 meters from his potential murderer for the bullet to miss its target. An increase of just 50 percent in James Bond's distance from the flight path can thus only be offset by a 16-fold increase between victim and perpetrator. Once again, the reason is the strong distance dependence of the magnetic field. In principle, it is therefore possible to protect a victim with a magnetic watch. However, 007 has to venture very close to the line of fire, in fact, within about one meter.

But there is still one thing that is extremely important when using a magnetic watch to protect people from pistol bullets. Because of Lenz's rule, the magnetic field of a coil does not arise immediately, but builds up slowly with a certain time delay.[9] The maximum field strength is not reached until a certain time after the watch is switched on. For the parameters of the best magnetic watch that we constructed earlier, this results in 90 percent of the maximum field strength being reached only after about 3.5 seconds. And that's exactly what we see in the movie scene. Only after a delay of about three seconds after switching on the magnetic watch does the spoon on M's saucer start to fly. That's the way it has to be. The spoon should not fly away immediately—that would be forbidden by physics.[10] Obviously, the watch we designed has exactly the same characteristics as the magnetic watch from the film *Live and Let Die*. These characteristics are summarized again in the table in Figure 6.8.

If Bond now wants to use this watch to save people who are in the line of fire, then he could in principle do so with the magnetic watch, but he must not only be very close to the line of fire, he must also switch on his watch three seconds before shooting. This certainly requires once again all the best attributes of a double-O agent.

Nevertheless, we can conclude that James Bond's magnetic watch would be technically feasible—purely in principle. However, the required materials are at the absolute limit of what is technically possible today, and some optimism would be required if anyone really wanted to build this watch. However, even if it were to work, its use would be severely restricted. Although it could attract teaspoons over a distance of 1.2 meters, heavier objects could certainly

---

[9] Lenz's rule states that the induction current always counteracts its cause. In the case of the magnetic watch, the cause is the electric current flow, and the magnetic field corresponds to the induction current. The build-up of the magnetic field thus temporarily inhibits the flow of current through the watch, and this in turn causes the magnetic field to build up more slowly. The current flow through the coil thus obstructs itself, which is why we speak of the phenomenon of self-induction.

[10] This scene therefore belongs in every physics textbook as a prime example of Lenz's rule. The makers of the film certainly didn't know this rule. It's probably pure coincidence that three seconds pass before the spoon flies off.

not be attracted over even greater distances owing to the strong distance dependence of the attraction. It would also be difficult to use the magnetic watch for personal protection. Here one must not be much further than one meter from the line of fire in order to significantly deflect the trajectory of a projectile. M's concerns expressed in the film about the British taxpayer are therefore not entirely unfounded.[11]

After M and Moneypenny have left, the top agent turns to a beautiful Italian, whom he has hidden in the closet until then, to conceal her from his superior. With an intimate embrace 007 opens the zipper of her dress with his magnet watch, whereupon the lady comments with a sigh *"Ohhh. Such a delicate touch", and* 007 replies: *"Sheer magnetism, darling."*

It goes without saying that the zipper of the lady's dress can easily be opened with this great magnetic watch, provided the zipper is made of iron or steel. The distance between the magnetic watch and the zipper in the scene is less than ten centimeters. The force available for opening is thus $12^7 = 35$ million times stronger than the force on the spoon mentioned earlier. That should certainly suffice to open a zipper. Perhaps the attraction would be so strong that she would now even be ripped out of her dress. However, opening the zipper of the Italian lady's dress would also work with the weaker watch, in which only 100 turns are wound on an iron core.[12] So Bond should rather have said *"It's the distance, darling."*

Details for Know-It-Alls

Let us now explain in more detail why the magnetic attraction decreases so much with distance and discuss the quantities it depends on. The coil in the watch is a magnetic dipole.[13] The magnetic field of a dipole decreases with the third power of the distance. This means that at ten times the distance the field has decreased to $1/10^3 = 1/1000$. Furthermore, the magnetic field is propor-

---

[11] For the optimal watch, we did not even discuss what energy the watch batteries would have to store. This energy would naturally be the essential reason why such a watch magnet will probably never be realized. Even 5 amps is far too high a current for a small watch battery.

[12] Experiments by the authors in Dortmund department stores show that a good zipper can be opened if a weight of about 100 grams is attached to the eyelet. This force has been used for the calculation. Unfortunately, the test series was brought to an end by salespeople who considered that one must also bring along a lady to buy a sinfully expensive lady's dress …

[13] Electrical dipoles are better known because they are formed when a positive charge and a negative charge are at a certain distance from each other. However, there are no magnetic charges, so magnetic dipoles are somewhat more complicated than electrical ones. For example, a rod magnet is a magnetic dipole whose field lines everyone has probably seen before in high school. Since there are no magnetic monopoles, magnetic fields are always at least dipole fields.

tional to the current flowing through the coil, the number of turns in the coil, the cross-sectional area of the coil, and a constant µ which depends on the material of the core of the coil. The latter is the magnetic permeability. It has the values µ = 5000 for iron and µ = 500,000 for amorphous metals such as metallic glasses made of ferromagnetic materials. If µ is much greater than 1, the material is said to be ferromagnetic. For most substances, however, µ is approximately equal to one.[14]

The iron spoon is then transformed by the magnetic field of the watch itself into a magnetic dipole, whose field strength depends on the external magnetic field. This is referred to as an induced dipole. Since the external magnetic field of the watch itself decreases as the third power of the distance, it follows that the resulting force of attraction decreases at least as the sixth power. Furthermore, the exact physical derivation shows that the attraction force on the spoon does not depend on the magnetic field itself, but on the *spatial rate of change* of the field.[15] The result is that the magnetic field decreases as the seventh power of the distance, i.e., it has already dropped to a ten-millionth at ten times the distance.

The detailed calculation of the current needed for the magnetic watch to attract the spoon is quite complicated and would require too much space here. In addition to the distance dependence already discussed, it is clear that the force of attraction depends on the square of the coil current I, the square of the number of turns N, the square of the material constant µ, and the square of the cross-sectional area of the coil, i.e., the fourth power of the watch diameter d. This follows from the same arguments which were already discussed in connection with the distance dependence of the original dipole field, since the induced dipole moment is proportional to the applied field. The magnetic force of attraction F_{mag} therefore satisfies.

$$F_{mag} \propto I^2 \cdot \mu^2 \cdot N^2 \cdot d^4 \,/\, R^7$$

where R is the distance between the watch and the spoon, and the sign $\propto$ means that the force is proportional to the quantities on the right-hand side of the equation and that a constant factor is still missing from this formula. If the spoon is to be attracted, the magnetic force of attraction must be greater than the weight of the spoon, i.e.,

---

[14] If µ is slightly greater than one, the substance is said to be paramagnetic. If µ is slightly less than one, the material is diamagnetic.

[15] For experts, this change is determined by differentiating the distance law with respect to the position. And the derivative of the function $1/R^6$ is proportional to $1/R^7$.

$$F_{\text{mag}} > m \cdot g$$

where m is the mass of the spoon and $g = 9.81$ m/s^2 is the acceleration due to gravity. If we now insert the magnetic attraction force in this inequality, then a condition can be specified for the coil current I which is needed to attract the spoon. The exact calculation finally results in the complicated formula

$$I^2 > 32 \cdot g \cdot \cdot \rho \cdot \mu_{\text{spoon}} \cdot R^7 / \left(3 \cdot \mu_0 \cdot \left(\mu_{\text{spoon}} - 1 \right) \cdot N^2 \cdot d^4 \cdot \mu^2 \right)$$

where ρ is the density of the spoon material, hence 7.5 g/cm^3 in the case of iron, μ_{spoon} is the magnetic permeability of the spoon material, which has the value 5000 for iron and is defined analogously to the constant μ for the coil core, and μ_0 is the magnetic field constant with the value $\mu_0 = 4.\pi.10^{-7}$ Vs/Am. This formula was used to calculate the current values in the text for the various configurations of the magnetic watch.

If the formula for the magnetic attraction force is examined more carefully, we see that, in addition to the strong distance dependence, there is also a strong dependence on the diameter of the magnetic watch. The above formula implies $F_{\text{mag}} \propto d^4$.

For the magnetic watch, it is therefore not only problematic that the spoon is at the relatively large distance of 1.2 m, but also that the watch has only a relatively small diameter. However, with a large electromagnet, such as those found in junkyards, it would be easy to attract a spoon from over a meter away. The diameter of such an electromagnet is about fifty times greater than the diameter of the magnetic watch. The attraction is then $50^4 = 6,250,000$ times stronger.

In the film *You Only Live Twice* from 1967, James Bond is pursued by Japanese villains in a car. However, the Japanese secret service comes to 007's aid with a helicopter carrying a large magnet. This magnet is lowered onto the roof of the car and the car is lifted off the road, to James Bond's delight. This scene is of course realistic, because on the one hand the distance is small because the magnet is actually in contact with the roof, and on the other hand the diameter of the magnet is large. Both effects dramatically increase the magnetic attraction.

The calculation of the temperature to which the magnetic watch would be heated is carried out in the same way as in the *Goldfinger* chapter on the melting of metals by laser beams. An energy supply E can easily be converted into a temperature change ΔT using the formula.[16]

---

[16] We should also ask whether this formula applies at all at such high temperatures as those occurring in the text. For reasons of simplicity, the heat of fusion and heat of evaporation were not taken into account.

$$E = c_{\text{watch}} \cdot m_{\text{watch}} \cdot \Delta T$$

The specific heat c_{watch} is a known constant for any given material and m_{watch} is the mass of the watch. The energy E comes from the battery of the watch and is assumed to be given by

$$E = U \cdot I \cdot t = R \cdot I^2 \cdot t.$$

where U is the voltage of the watch battery, R is the electrical resistance of the magnetic coil in the watch, and t is the time during which the watch magnet operates. The electrical resistance R of a wire of length L and cross-section A can be calculated using the following equation:

$$R = \rho_{\text{sp}} \cdot L / A = \rho_{\text{sp}} \cdot \pi \cdot d \cdot N / A$$

where ρ_{sp} is the specific resistance of the coil material, i.e., the specific resistance of copper. This number is also a known constant. The cross-sectional area A and length L can be calculated from the geometry of the coil. Putting all this together, we deduce an expression for the temperature change ΔT which the watch experiences for the given current flow:

$$\Delta T = \pi \cdot \rho_{\text{sp}} \cdot d \cdot N \cdot I^2 \cdot t / \left(A \cdot c_{\text{watch}} \cdot m_{\text{watch}} \right)$$

We used this formula to calculate the temperatures indicated earlier, assuming suitable values for c_{watch} and m_{watch}.

A Watch, a Steel Cable, and Lots of Fantastic Physics

The mysterious Elektra King wants to have a pipeline built from Azerbaijan to the Mediterranean in the 1999 film *The World Is Not Enough*. But there are already three Russian pipelines in the north that reach as far as the Black Sea. From there, the oil is transported by ship. Elektra King is kidnapped and brainwashed by the terrorist Renard, who wants to seize control of the world's oil supply. When Renard has the nuclear scientist Dr. Arkov killed, Elektra King's bodyguard Davidov takes over his duties. James Bond sees his chance to find out what Renard is up to and pretends to be Davidov. He flies to Kazakhstan, where plutonium is being removed from old bombs in a dismantling bunker. There James Bond meets Dr. Christmas Jones, who unmasks him without realising what is going on. There is a shoot-out between 007 and Renard's men. James Bond and Dr. Jones are forced to jump into a pit lined with tiles. Renard closes all the doors—and there seems to be no way out for 007 and the beautiful atomic physicist. But James Bond won't let the bad guy get away with it. This is where his special watch with integrated hook and launch mechanism comes into play: the hook whizzes through the air at lightning speed on a long, thin rope 15 meters long and penetrates with great force into a girder which is located on a crane above the pit. James Bond is pulled up jerkily and lifted to freedom within two seconds, just manages to get through the closing gates and pursues Renard to the elevator. At the end of the film, 007 is able to hunt down Renard and Elektra King and prevent a nuclear submarine from exploding in the Bosporus.

The first question that arises is whether the hook can penetrate so deeply into the girder that it can withstand James Bond's weight. It is obviously a heavily rusted steel girder. The rope from the watch seems to be a conventional steel rope. For the exact analysis the total height of the room is needed. This is determined by counting the individual crane elements. One crane element corresponds in length to four tiles attached to the wall of the pit and a tile is about 30 centimeters across. Thus the total height of the room for eleven crane elements is approx. 13 meters. So when Bond's height of 1.83 meters is subtracted from the total height and the hook is considered to fly at a slight angle, we find that it does indeed shoot a total distance of about 15 meters into the air. The speed of the hook shooting upwards is determined by the distance travelled and its flight time. With a measured flight time of 1.5 seconds, the speed of the hook is 10 meters per second, i.e., 36 kilometers per hour. That's not actually very much for a hook that obviously penetrates the steel girder like a hot knife through butter.

But how deep does it really go? When deforming a material such as steel, two different ranges are basically considered: the elastic and the plastic range. In the elastic range, the steel girder deforms, but returns to its initial position as soon as the load is removed. In the plastic range, on the other hand, the steel girder is permanently deformed. It is not possible to say exactly which range is involved in the film scene. The forces expected for the elastic range are smaller than those expected for the plastic one, which is why only this range should be considered. A very simple model is used, which was actually developed for the elongation of bars. In order to transfer the model to the case of the hook, a bar-shaped area is imagined to be cut out of the steel girder (in thought alone). It is precisely this rod-shaped area of the material that is deformed by the impact of the hook. We consider that the area outside is not relevant to our considerations.[17]

The first step is to calculate the force needed for the hook to penetrate by five millimeters. This depends on the size of the impact surface and the modulus of elasticity of steel. The modulus of elasticity indicates how strongly a body can be compressed with a certain force. It is a known constant specific to each material. Given the modulus of elasticity of steel and the dimensions of the tip of the hook, it is found that a force of 14,000 newtons would be needed to achieve a penetration depth of five millimeters. That's a huge power. By way of comparison, to drive a nail five millimeters deep into wood with a hammer, only a force of 15 newtons is required. So almost a thousand times more force is needed to deform the steel by just five millimeters. The whole thing becomes even clearer when the required force is converted into the speed that the hook would have to have to penetrate this far into the steel. This would be about 430 kilometers per hour[18]!

The previously calculated launch speed of the hook at a maximum of 36 kilometers per hour is therefore in no case sufficient to make even the smallest dent in the steel. Strictly speaking, however, the conditions are even less favourable: the hook must not only compress the rod-shaped area when it penetrates, but the steel must also be pushed outwards, away from the incoming tip of the hook. This shear force can be calculated in a similar way to the force used to push the steel beam in. The result is that 62,000 newtons are required this time. Converted into the speed of the hook, the result is such an unrealistically high value that it would be better not to specify it here.

So is it completely impossible for the hook to penetrate the girder, as shown in the film? For a steel girder, this question can be answered with a clear "yes". However, if we imagine that it is a wooden girder that merely looks like a steel girder, the results become more realistic. Although calculations still have to be carried out as before, the modulus of elasticity is now many times lower. This is why speed of only 14 kilometers per hour is required for the hook to penetrate five millimeters into the beam. The launch speed of the hook is certainly a little higher than that. Five millimeters and possibly a little more penetration depth would then be possible without problem. If, however, the whole length of the hook is required to penetrate the girder so that 007 can

---

[17] In reality, of course, this will not be exactly right. However, if the correct deformation were taken into account, the calculation would be much more complicated. Thus, we try to make a rough estimate of the circumstances by simple means.

[18] Note that we only calculate the speed for a five millimeter deep penetration of the hook into the steel beam. But that wouldn't be deep enough to carry James Bond's sporty 76 kilograms.

hang securely from it, a speed of at least 250 kilometers per hour would be required, which shows that this is hardly possible even for a wooden girder.

Suppose Q has solved this problem for 007, and the hook is shot out of his watch so quickly that it penetrates completely into the wooden girder. Can Bond really be pulled up by a strong electric motor on the steel cable as easily as shown in the film? For this we have to calculate the force James Bond's arm has to withstand when he is pulled up jerkily. The watch on his wrist carries him up about 15 meters after the hook has been fired, and he does not seem to feel any pain. 007 obviously does not even have a tense shoulder joint after this action. This is somewhat surprising, because his arm, hanging on a rope, not only has to withstand his entire body weight, but also the force of the watch motor, which acts for a short time at the beginning. Exactly how great would this force be?

This can be calculated relatively easily using Newton's second law. Since James Bond weighs 76 kilograms, the mass that is accelerated is known exactly. The acceleration can be calculated from the speed reached by the secret agent when he begins to move evenly. In fact, a distance of 15 meters is travelled and the time required for this is two seconds, implying a speed of 27 kilometers per hour. James Bond is accelerated from zero to this speed within 0.2 seconds, as a single frame analysis of the corresponding film scene shows. Thus, his acceleration is 37.5 meters per second squared or 3.8 g. The force acting on his arm is thus 2850 newtons, which would be the weight force on a mass of 285 kilograms.

But that's not all. We still have to add James Bond's weight to the balance of forces as shown in Figure 6.9. Even for a very slow acceleration, his arm would still have to withstand at least his own body weight. Thus, not only the calculated accelerating force, but also the body weight acts on 007's arm.

The sum of this force and the previously calculated accelerating force of the engine is the actual total force of about 3600 newtons acting on James Bond's arm. This means that his arm must withstand a force corresponding to the weight of a mass of about 360 kilograms.

So the arm of the top agent is pulled up jerkily with this very large force. What would happen? Let's make this clear with a comparison that everyone knows: the use of toilet paper. First we slowly pull the roll because we need the paper. The speed of the paper changes slowly, and a small force is acting. This unrolls the paper. Then we pull jerkily, so that the paper is accelerated strongly. Correspondingly a large force is acting, and the paper tears at its weakest point, i.e., where it is perforated. If James Bond were such a roll of toilet paper, the perforated area would probably be at the height of his shoulder. It is therefore very likely that our top agent's arm would be torn off and pulled

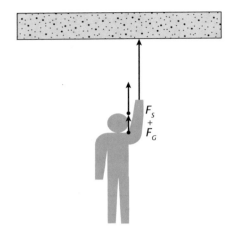

Figure 6.9 The forces acting on the rope and James Bond's arm while he is being pulled up with the rope in the watch. At the same time, the arm and the rope must not only withstand the accelerating force F_S, which is generated jerkily by the watch motor, but also the weight force F_G

up by the rope, leaving the rest of his body motionless on the ground due to its inertia …

Of course, the question arises once again of what energy would be needed to accelerate James Bond and pull him a distance of 15 meters further up. Could a normal watch battery do that?

His kinetic energy is approximately 2100 joules when his mass of 76 kilograms moves upwards at 7.5 meters per second on the rope. When he arrives at the top, he has gained in addition a potential energy of about 11,000 joules.[19] If these two results are combined, the total energy required is about 13,000 joules. This is not a very large amount of energy. The nutritional value of a single chocolate bar is many times greater.[20] However, this amount of energy must be delivered by the watch in a very short time. A normal button battery, as used in watches, has an operating voltage of three volts and a charge of 210 milliampere hours. In the short time James Bond is pulled up on the rope, it can only deliver 1.3 joules of energy. This means that the secret agent would need 10,000 such button cells to free himself with the help of his

---

[19] Potential energy depends only on the height at which a body is located. When a ball is rolled up a mountain, energy can be stored in this way. When it rolls down again, this stored energy is released again in the form of kinetic energy.

[20] For example, the calorific value, i.e., the energy content, of a Snickers bar is 2000 kilojoules, or 2 million joules!

watch. Instead of the 10,000 button cells, a super battery from Q would of course be sufficient.[21]

Now we have to take a closer look at the rope. Can a rope that has to withstand a weight of around 360 kilograms fit into such a stylish watch? A steel cable intended to carry 360 kilograms must be at least 2.2 millimeters thick. In order for 15 meters of such a rope to fit into the watch, it must have a sufficiently large diameter. Since the rope is shot out of the watch without any appreciable delay, we assume that it is rolled up without overlaps, similar to a liquorice wheel. Otherwise, the likelihood of it getting entangled would be far too great. Given the thickness and length of the rope, the diameter of the watch would be at least 20.5 centimeters. It would therefore be much larger than the one shown in the film, which has a diameter of just under four centimeters. Conversely, if the real diameter of the watch is fixed at a maximum of four centimeters, it would have to be almost five centimeters thick, assuming that the rope is wound up like a reel of thread and fills the watch completely. That would be a rather thick watch, without even talking about the space required for the launcher …

James Bond's watch therefore has many characteristics that are not easy to explain. In particular, the speed that the hook must have to penetrate the steel girder makes it extremely difficult from a physical point of view. However, if it is a wooden girder made to look like a steel girder, if the watch is powered by a special battery, if the rope is made of a particularly tear-resistant material, and finally if James Bond has an extremely hard-wearing shoulder to absorb a jerk equivalent to the weight of 360 kilograms, then this watch would be nothing special.

Details for Know-It-Alls

The modulus of elasticity E is a constant specific to a given material. In the elastic range, the stress σ is given by $\sigma = E \cdot a$, where the "elongation" a is in this case the penetration depth as a fraction of the total length. The stress is the force per unit area applied to the steel girder. If a force F applies to an area A and compresses the steel beam of thickness D through a distance L, this force is then given by

$$F = E \cdot A \cdot L / D$$

---

[21] As we already know from the movie *Live and Let Die*, Q did indeed invent such a battery for the fantastic magnetic watch!

In the present case $L = 5$ mm, $A = 1$ mm^2, $D = 7$ cm, and $E = 2 \cdot 10^{11}$ N/m^2 for steel. This results in a force of 14,000 N. This force can be used to calculate the speed required by the hook to penetrate the girder. If the hook has kinetic energy $\frac{1}{2} \cdot m \cdot v^2$ and mass $m = 10$ g, we can equate this with the work that the force F must perform by pushing the steel surface over the penetration distance L. Then the square of the velocity v of the hook is given by

$$v^2 = 2 \cdot F \cdot L / m$$

The speed $v = 430$ km/h can be calculated from this by inserting the known numerical values.[22]

In order to calculate the energy E_{bat} that the watch batteries have to provide so that James Bond can be accelerated jerkily to a speed of $v = 27$ km/h and then pulled up a distance of $s = 15$ m, we must add together the energy used to accelerate his mass $m_{Bond} = 76$ kg to his final speed and the energy subsequently used to pull him up. This leads to

$$E_{bat} = \frac{1}{2} \cdot m_{Bond} \cdot v^2 + m_{Bond} \cdot g \cdot s$$

The first part is Bond's kinetic energy and the second part is the increase in his potential energy, while $g = 9.81$ m/s^2 is the acceleration due to gravity, as before. Substituting in the numbers results in an energy of about $E_{Bat} = 13,000$ J. If the button cell of the watch supplies a current of 210 milliamperes at an operating voltage of 3 V, this results in an energy of $E_{button} = 3 \cdot 0.210 \cdot 2 = 1.3$ J, which the battery would be able to deliver during the 2 s in which James Bond is hanging from the rope. So 10,000 such button cell batteries would actually be needed.

From the force $F = 3600$ N which the rope must withstand, its diameter d can be calculated in a rather complicated way. The following applies:

$$F = \frac{1}{4} \cdot \pi \cdot f \cdot B \cdot k \cdot d^2 \approx 700 \, \text{N} / \text{mm}^2 \cdot d^2 ,$$

where the filling factor f is the proportion of the steel cross-section to the total cross-section, B is the hardness, k is the stranding factor, and $\pi = 3.14$. These

---

[22] However, it must be stressed here that these considerations are very much simplified. In reality, the penetration of a hook into a steel girder is a highly complex process that cannot be described so simply. The calculation given is only intended to illustrate the orders of magnitude of the forces and speeds that must be involved.

figures can be found in data tables for steel cables, hence the value of 700 N/ mm² given here. Since we know the force which the steel rope must withstand when James Bond is pulled up jerkily *is* at least F = 3600 N, it follows from the above formula that the rope must have a thickness of d = 2.2 mm. From the thickness d and the length L_{rope} *of* the rope, the diameter U of the watch follows immediately by equating the surface which the rope takes up lengthwise with the circular surface which is created when the it is tightly wound up in a spiral. This yields

$$U^2 = 4 \cdot d \cdot L_{\text{rope}} / \pi.$$

With the known numerical values, we find U = 20.5 cm for the diameter of the watch. However, if the rope is wound up in a reel, we can calculate the height H the watch would have to have if its diameter were fixed at U = 4 cm, then a similar consideration implies

$$H^2 = 4 \cdot d^2 \cdot L_{\text{rope}} / (\pi \cdot U).$$

Inserting the numbers now implies H = 4.8 cm, i.e., almost 5 cm.

Sunglasses à la Bond

In the 1985 film *A View to a Kill*, James Bond visits Chantilly Castle, about 50 kilometers northeast of Paris. During his search for information about the mysterious Max Zorin, he is denied access to a room on the ground floor. James Bond goes outside and looks for a window not covered by curtains. It is a beautiful summer day, the sun is shining and it is difficult for 007 to look into the room due to the strong reflections. But the top agent is very well equipped and puts on his very special sunglasses. He rotates the glasses and for a certain position can watch the events in the room undisturbed and without a fuss. He does not have to press his nose against the pane because he is not hindered by the bright reflections.

How is that possible? Can a bright reflection simply be "rotated away"? To understand this, we must take a closer look at the properties of reflected light. To do so, we must consider light as an electromagnetic wave. Strictly speaking, light is a transverse wave, i.e., the electromagnetic field oscillates perpen-

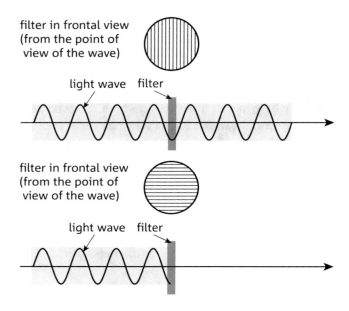

Figure 6.10   If the wave oscillates in the direction of the bars of the polarization filter, it can pass through unhindered (top). If the bars are perpendicular to the direction of vibration of the wave, it is completely absorbed (bottom)

dicular to its direction of propagation.[23] This direction of oscillation of the field is called the polarization direction of the wave, or simply the polarization.

A polarization filter, or polarizer for short, can be imagined as a grid with vertical bars. If the waves striking the polarizer oscillate in the same direction as the rods, they can pass through the grid unhindered. However, if the bars are perpendicular to the plane of vibration of the wave, it cannot pass through the polarization filter. The wave is absorbed. This is illustrated schematically in Figure 6.10. For electromagnetic waves with longer wavelengths, grids of metal rods are actually used as polarizers. But light has wavelengths between 400 and 700 nanometers, so a polarization filter for light must have grid spacings of this size. That is why polarization filters are made of herapathite, for example.[24] This substance comprises long iodine chains which act like microscopic lattice bars and thus give the material its polarizing properties. Thin films can be made from herapathite and used as polarizing filters for cameras.

---

[23] There are also longitudinal waves in which the wave oscillation takes place parallel to the direction of propagation. Sound waves propagate as pressure fluctuations in the air and are in fact longitudinal waves.

[24] The chemical formula of herapathite is $C_{80}H_{104}I_6N_8O_{20}S_3$

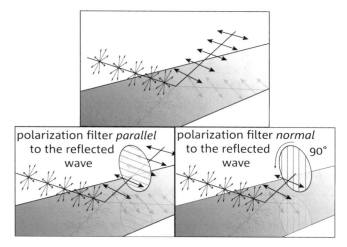

Figure 6.11   Upper: When unpolarized light is reflected from a surface, it is (partially) polarized, oscillating mainly in one direction. Lower: This reflected radiation can now be suppressed with a polarization filter. On the left we see the position in which the reflected light can pass completely unhindered through the polarization filter; on the right it is blocked. The two positions differ by an angle of 90 degrees

Sunlight is unpolarized. The individual light waves thus oscillate in a disordered manner in all different directions. This is due to the process by which the light is produced, in which no direction of radiation is preferred.[25] However, this no longer applies to light reflected from a surface. After reflection from a surface, a direction is preferred, and the reflected light is at least partially polarized.[26] It therefore oscillates mainly in one direction and can indeed be suppressed with a polarization filter, as can be seen schematically in Figure 6.11.

To achieve this, James Bond must adjust his polarization filter so that it is perpendicular to the plane of vibration of the interfering reflected light as shown in Figure 6.11 (bottom right). He rotates the filter until the reflections almost disappear. If you look closely, you can see that James Bond twists the polarization filters of his glasses by 90 degrees. This is exactly the angle between the two extreme positions shown in Figure 6.11. Bond actually looks for the optimal setting for reflection suppression. Here the representation is thus perfectly realistic!

---

[25] Laser light is completely polarized because of the way it is generated. It constitutes an exception. As a rule, the light produced by light bulbs and other kinds of lamp, or indeed in nature is always unpolarized.

[26] The exact reason is that so-called s- and p-polarized light is reflected differently.

An exact analysis shows that the light is not polarized equally at all reflection angles. The angle between James Bond's viewing direction and the window pane should be between 50 and 60 degrees. This range contains the so-called Brewster angle, at which the reflected light is completely polarized.[27] This is where it can best be filtered away. If you take a closer look at the movie scene, you will first see two ladies standing near the window. James Bond greets them in a friendly way, makes a left turn, walks towards the window, and stops at some distance. In the scene we can see that 007 remains at each point at a greater angle than 45 degrees to the pane and at a smaller angle than 65 degrees when he looks through the window with his polarizing glasses. Once again, James Bond must have enjoyed an excellent education in physics with the British secret service, because he obviously knows the angle at which he has to position himself in front of the window pane in order for his polarizing glasses to function optimally. Most likely, these polarizing glasses actually work. Anyone could easily build such a pair of glasses from the polarization filters of photographic equipment, because these filters work according to the same principle.[28]

"I See Something You Don't See"

In the movie *The World Is Not Enough*, in search of his old friend Valentin Zukovsky, James Bond enters his Casino. While he's looking at one of the guards, 007 puts on his special glasses. This allows the secret agent to see through clothing—firearms, knives, and women's underwear worn underneath shine in a blueish-white light. The interior and the people themselves look quite normal, except for a light blue veil. After he has found Zukovsky's most heavily armed confidant, he has a message delivered to him and first of all allows himself a drink—vodka martini, of course shaken and not stirred.

But how exactly does this pair of miracle glasses work? They seem to be X-ray specs. On the Internet one can even buy "X-ray glasses" that look exactly like James Bond's X-ray glasses.[29] Let us now analyse this step by step. First of all, it should be noted that it is probably not possible to X-ray any objects with X-ray glasses. They are not an active element from which rays emanate, but

---

[27] The Brewster angle depends on the refractive index of the material comprising the given reflective surface. For glass, with a refractive index of 1.5, it is 56.3 degrees. Experts know that this corresponds to the arc tangent of 1.5.

[28] Anglers know that you can even buy polarization glasses with a fixed setting. This eliminates annoying reflections of sunlight and fish can be seen better in the water.

[29] Such "X-ray glasses" were available for a long time at www.asseenonscreen.com for a few pounds.

can only be a passive device, like an X-ray detector. We have already seen this in the analysis of the film *Moonraker*. There, 007 also had a small detector and was able to crack the lock of the secret safe. An object must be located between the X-ray source and the detector to be X-rayed.

We therefore assume that James Bond only carries a detector. The X-ray source is located somewhere behind the walls of the casino. If only one X-ray source were used for illumination, the X-ray shadow cast by the group of people would be larger than the group itself and thus much larger than the spectacle lenses. So that couldn't work. However, it would be possible for James Bond to hide a detector in his jacket pocket that measures X-rays in a direction-selective way. This would work if the detector were in a lead box with a small hole, while Q or another helper "floods" the casino with X-rays from all directions. The detector would scan all directions one after the other and measure the X-rays coming from there. Like a television set, a computer could then compile an image point by point, and this could then be projected directly onto the secret agent's spectacle lenses. So that's how it could work, and the computer could of course be hidden in the watch on his wrist.

But what exactly does it mean to flood a casino with X-rays? There should be an X-ray tube behind each pixel. If details of one centimeter in size are to be detected, then one tube per square centimeter of wall would also be required, i.e., 10,000 tubes per square meter. If each X-ray tube needs an output of 100 kilowatts, the total output would be one gigawatt per square meter of wall![30]

Even if we forget the enormous costs and also the expenditure required to cool the X-ray tubes,[31] the question still arises as to what could cause so much X-ray radiation. Water, the main component of the human body, would heat up so much as a result of the absorption of radiation that it would begin to boil after about 16 seconds. The same would of course apply to James Bond's vodka martini. However, visitors to the casino would not notice much more, because the radiation exposure would be a fabulous 19,600 gray per second. After just three thousandths of a second, the dose of 50 gray, which is immediately fatal for humans, would be reached. Even though James Bond has been toughened to the extreme by his extensive training with the British secret service, he could not be responsible for "flooding" a casino with X-rays in this way just to look under other people's clothes with X-ray glasses.

---

[30] A tube of at least 100 kilowatt would be needed to X-ray a complete human being at this distance. The radiation from each tube must cover the entire area of the casino. Remember that 1 gigawatt is the power generated by a medium-sized coal or nuclear power plant!

[31] Secret services usually have unlimited budgets. No surveillance has yet failed for financial reasons …

Figure 6.12 Detailed analysis of the images James Bond sees through his glasses *in* Valentin Zukovsky's casino in the film *The World Is Not Enough*. In the left-hand picture, the woman's pistol is clearly visible, … but when the man passes in front of her in the right-hand picture, she disappears. **However, an X-ray shadow cannot simply be concealed!**

So we need to think this over from another angle. In principle, James Bond could also carry the X-ray source with him and the walls of the casino could have been secretly equipped with X-ray detectors by the British MI6. The signals from the detectors could then be transmitted by radio to James Bond's glasses, where an image would be assembled from the data and projected onto the lenses. Since this solution requires only a single tube, the radiation exposure is also quite small. But the problem of waste heat is somewhat tricky again. If there were no cooling, even a good X-ray tube would only last about ten seconds at full load before its anode would begin to melt. But the casino scene in *The World is Not Enough* lasts a full 46 seconds. James Bond would be unlikely to drag along a cooling circuit for the X-ray source. However, by pausing after taking an X-ray image, he could stretch the time out for long enough. For a regular film with 25 frames per second, there would still be nine thousandths of a second of exposure time left, which is a bit tight, but not impossible. In principle, James Bond could cover the power requirements for the X-ray source using lithium-ion batteries weighing a good 1.6 kilograms.[32]

X-ray irradiation reveals all absorbers between the source and detector. A passing individual could never conceal another person's weapon, in contrast to what is shown in Figure 6.12. But the situation is even worse: the bones of all the people, the tubes in the wall, and all other objects could always be seen on a conventional X-ray! The images of real X-ray glasses would therefore look

---

[32] Such batteries are also used in notebooks and smartphones.

completely different from what James Bond gets to see in the film. It is therefore highly probable that a different technology is actually being used.

However, there is another method that could overcome some of these problems and which would also be more suitable for X-ray glasses: X-ray backscatter technology. This method does not use the penetrating part of the X-ray radiation but rather the weak part scattered back to the source by so-called Compton scattering.[33] James Bond could then carry the source as well as the detector with him, and the building work that would have to be carried out beforehand in the casino by the British secret service would no longer be necessary. Since the scattering takes place in all directions, the object must be scanned bit by bit with a narrow X-ray beam and an image built up from this. However, large-area detectors can also be used for this. Then, although the backscattered portion is only 0.8 percent, even a small 1.4 kilowatt X-ray tube would suffice for this technology. The radiation exposure would also be relatively low. People scanners that work according to this principle have already been used at London Heathrow Airport and Sky Harbour International Airport in Phoenix/Arizona: they are, however, the size of a telephone box.[34] Although the X-rays used here penetrate clothing, they only penetrate the body to a maximum depth of two millimeters and are then reflected. Since the scattering signal is particularly pronounced for light elements which are mainly found in organic substances, not only metals but also non-metallic weapons, explosives, and drugs can be detected. Figure 6.13 illustrates the principle once again.

With the Compton backscatter technique, there is also a delicate problem that would not have bothered James Bond—on the contrary. Since no X-ray resistant underwear has been invented so far, everyone will be seen completely naked on a backscatter image. Passengers at London Heathrow were only moderately enthusiastic when they heard about this. In the end, the company which produces the backscatter scanners for airports had to develop software that distorts the images to such an extent that only shadowy outlines can be seen. Perhaps the brightly lit lingerie that James Bond can see through his X-ray glasses is just a software effect that Q has built in so that 007 won't be too distracted from his work.

---

[33] Named after Arthur H. Compton, who, in 1923, pioneered experiments on the scattering of X-ray radiation on atoms. In 1927, Compton was awarded the Nobel Prize for Physics.

[34] For younger readers: before the invention of the mobile phone or smartphone, there were small public rooms everywhere with a telephone, in which there was only room for one person. From there you could call other people. Such rooms were called phone booths or telephone boxes.

Figure 6.13 In the Compton backscattering method, the object is scanned point by point with a very narrow X-ray beam and the weak, backscattered radiation is measured (indicated schematically on the left). In the right-hand picture, a car was scanned with this backscattering method. The metal parts hardly scatter the radiation, while organic substances clearly emerge. The white packs are hidden cocaine

If we move away from the idea that X-ray glasses must be involved, there would be another technology that would make the film scene very realistic.[35] Terahertz technology uses electromagnetic radiation with frequencies in the range of one terahertz.[36] This sounds like a lot, but it is not much compared to light or X-rays. The wavelength of the terahertz radiation is correspondingly large, in fact in the range up to one millimeter. By way of comparison, if the wavelength of X-rays were increased to the thickness of a human hair, the wavelength of terahertz radiation would be as large as a detached house.

Terahertz radiation is therefore much less energetic and, unlike X-rays, cannot damage our health. The possibilities for applications are similar to those with X-ray backscatter technology. There are also person scanners, as shown in Figure 6.14. Here, too, the radiation easily penetrates clothing, but is strongly absorbed by the water or other substances in body tissue. Unfortunately, the underwear problem also exists here. In fact, a particular terahertz source is not always needed, because the natural radiation of the body is already sufficient. A further advantage of terahertz radiation is that it can be focused and sharply imaged by lenses and mirrors just like light, while X-rays cannot. This means that real terahertz cameras can be built that function in a similar way to thermal imaging cameras. Because the image does not have to be scanned point-by-point with this method, films can also be recorded in real time. So it looks like terahertz glasses are exactly what James Bond probably used here!

---

[35] The name "X-ray glasses" is never mentioned in the film itself, but is mentioned in many accompanying books.

[36] One terahertz (1 THz) corresponds to 1,000,000,000,000 oscillations of the electromagnetic field per second.

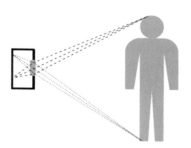

Figure 6.14 Terahertz radiation can be focused through a lens like light and imaged on a detector (left). One either uses the natural radiation emitted by humans or illuminates with artificial sources. On the right is an image taken at 0.1 terahertz. The knife hidden in the newspaper is clearly visible

Details for Know-It-Alls

X-rays, light, and terahertz radiation are electromagnetic waves. Like every wave, they have a wavelength and a frequency. The wavelength is the distance between two wave crests. The frequency indicates the number of wave crests passing an observer per unit time. If these two quantities are multiplied together, the propagation speed of the wave is obtained. This speed is the same as the speed of light for all electromagnetic waves. At almost 300,000 km/s, it is a natural universal constant.[37] The longer the wavelength, the lower the frequency must be. If the wavelength is very short compared to the dimensions of obstacles, then the electromagnetic waves propagate quasi-linearly.[38] One then speaks of light or X-rays. High-frequency radiation is more energetic than low-frequency radiation. It can therefore trigger more energy-rich processes. X-rays are very high-energy radiation. Only gamma radiation, which is generated during radioactive decay, is richer in energy. This is why the penetration capacity of X-rays is so great. The overview in Figure 6.15 shows the different types of electromagnetic waves.

It is now possible to calculate how much radiation the guests in the casino would absorb if the room were X-rayed, i.e., "flooded" with X-rays. The power absorbed in a thin layer of thickness Δx is $P = \Delta I \cdot A$, where ΔI is the attenuation of the X-rays and A is the irradiated area. The mass of the thin layer is $m = \Delta x \cdot A \cdot \rho$, where ρ is the density of the material. If we now divide the first expression by the second, we obtain the absorbed power per unit mass with

---

[37] More precisely, this refers to the speed of light in a vacuum. In matter it may well be lower. The speed of light in a vacuum is the highest possible speed for anything.

[38] Otherwise, diffraction effects occur and the light will spread.

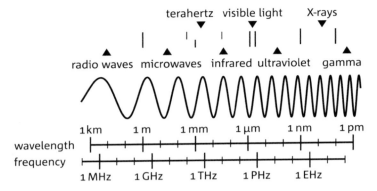

Figure 6.15 The electromagnetic spectrum. Each short vertical line on the scales indicates a frequency ten times higher than the last, or a wavelength ten times shorter. Radio waves are long wave electromagnetic waves. Higher frequencies are followed by microwaves and finally by infrared radiation. Because bodies radiate in this frequency range at room temperature, it is often referred to as heat radiation. Visible light, the light we can detect with our eyes, makes up only a small part of the electromagnetic spectrum. All the rest is invisible to the human eye. Ultraviolet radiation is higher frequency and therefore more energetic than visible light. Next come X-rays and the even more energetic gamma radiation. Terahertz radiation lies on the borderline between infrared and microwave radiation

the formula indicated in the chapter on *Moonraker*, and hence also the weakening of the intensity:

$$P / m = \Delta I \cdot \mu / \rho.$$

Putting in realistic values finally yields a result of 19,600 W/kg, which for X-rays is also the radiation dose in gray per second. By comparison, natural exposure is only about two to four thousandths of a gray per year. To heat water by 1 °C in one second, a power of 4180 W/kg is required. So it won't take long before the casino guests start sweating. For a body temperature of 37 °C, a time of $(4180/19{,}600) \cdot (100 - 37) = 13$ s will elapse before the water in the human body begins to boil.

This Backpack Is a Rocket

In the 1965 film *Thunderball*, James Bond escapes the villains from the Château d'Anet[39] estate at the beginning with the help of a jetpack. He first puts on a safety helmet and then flies out of the building in a few seconds. 007

---

[39] This castle lies 65 kilometers west of Paris and was built between 1547 and 1555 by order of Henry II, who gave it to his lover Diana of Poitiers as a present.

lands elegantly next to his Aston Martin, where a colleague is already waiting for him. Bond remarks: *"No well-dressed man should be without one."* Then they quickly stow the jetpack in the trunk and flee.

It is interesting that 007 still finds the time to put on his helmet. The reason for this is very simple: the jetpack really works! It was invented at the beginning of the 1960s and used here in the film. Since flying with a backpack can be quite dangerous, as we will see soon, the stuntman had to wear a safety helmet during his flight. To ensure that this would not attract attention, James Bond has to put on such a helmet before, although he is really short of time.

In the early 1960s, the American company Bell Textron developed a jetpack for the US Army known as the Rocket Belt.[40] This Rocket Belt used jet propulsion that worked with superheated steam. According to the manufacturer, a 20-second flight was possible with this model, which could reach a height of 18 meters. A maximum speed of 55 kilometers per hour and a maximum flight distance of 250 meters were possible. A computer program can be used to simulate the vertical flight of a person weighing 76 kilograms, followed by a gentle landing. The first step is to find out how the jetpack works. In principle, it works just like a real rocket. The downward ejection of fuel generates an upward thrust. Therefore, everything that we have already stated in the discussion of the film *Moonraker* can be applied here. In the Rocket Belt, however, hydrogen peroxide[41] was used instead of liquid hydrogen and oxygen as in the Space Shuttle. Hydrogen peroxide reacts under pressure to form a mixture of water vapour and oxygen. This mixture flows out from the nozzles of the jetpack at a temperature of about 750 degrees Celsius and thus generates the necessary thrust. The backpack used here held about 27 kilograms of hydrogen peroxide. Due to the escaping hot water vapour, a pilot must wear insulating clothing in order not to get burnt. With James Bond, we can assume that Q has given him a tailor-made suit that is both insulating and not flammable.

The technical data of the jetpack and James Bond's weight of 76 kilograms will be used in the program to simulate flight trajectories. The flight curve calculated in this way is shown in Figure 6.16. An altitude of about 25 meters and a flight time of about 20 seconds are reached. These values correspond very well with the values given by the manufacturer for the flight time and

---

[40] The Bell Rocket Belt was invented by Wendall F. Moore in 1961. In 1984, it became very well known when a stuntman with a jetpack arrived at the opening ceremony of the Olympic Games in Los Angeles and landed in the middle circle of the stadium. In the movie scene, the jetpack was flown by a stuntman. Only when James Bond was seen in close-up during takeoff and landing did Sean Connery hang on ropes hidden in the film.

[41] In contrast to water H_2O, hydrogen peroxide has the formula H_2O_2. Hydrogen peroxide is commonly used to bleach hair.

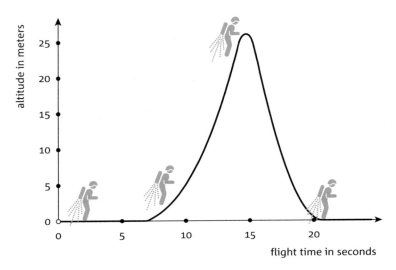

Figure 6.16 Calculated altitude of James Bond with a jetpack as a function of time. He can fly about 25 meters high with a total flight duration of 20 seconds. Then the fuel will be used up. We can also see that he doesn't take off right away

flight altitude. Here we can see that the pilot has to be very experienced and has to manage the fuel very well. If he's 25 meters up, the tank could already be empty and he'd just fall down. Therefore the jetpack can only be flown by very experienced pilots or stuntmen, and it is strongly recommended to wear a safety helmet.

The program for calculating flight paths is just as suitable for a Saturn V moon rocket as for a rocket backpack. The reason for this is that both function according to exactly the same principle. So James Bond's backpack is really a rocket!

Unfortunately, this means that its fuel consumption is also as high as that of a rocket. Flights lasting more than one minute would seem unrealistic with this principle. The fuel for a 20 second flight with the jetpack costs about 3000 dollars, which does not make it particularly attractive as a means of mass transportation either.

7

"Shaken, Not Stirred!"

Bartender: *"Can I get you something, sir?"*
Bond: *"Vodka martini—shaken, not stirred!"*
Bartender: *"I'm sorry, we don't serve alcohol."*
Bond: *"I'm really starting to love this place."*
Q: *"He'll have the prolytic digestive enzyme shake."*
Bond: *"Certainly."*
(Quote from the 24th Bond Film *Spectre*)

James Bond always takes his vodka martini shaken and never stirred. So far, he's ordered this drink 28 times in all the movies together. And he always warns the bartender: *"Shaken, not stirred!"* Why is he doing this? Is there a special reason for this, or is it just some kind of trademark? Fortunately, the recipe for the original secret agent cocktail can be found in Ian Fleming's 1953 novel *Casino Royale*:

Fleming's Original Recipe
"[Take] a dry Martini in a deep champagne goblet [...]. Three measures of Gordon's, one of vodka, half a measure of Kina Lillet. Shake it very well until it's ice-cold, then serve and add a large thin slice of lemon peel."[1]

The intensive shaking of the drink obviously lowers the temperature, as the drink then comes into better contact with the ice in the shaker. James Bond

---

[1] According to Bond, the cocktail tastes better if the vodka is made from grain, not potato. Kingsley Amis, in *The James Bond Dossier* (1965), suggests that Ian Fleming got it wrong and the cocktail should be made with Lillet rather than Kina Lillet, which he says would make it unpleasantly bitter.

© The Author(s) 2020
M. Tolan, J. Stolze, *Shaken, Not Stirred!*, Science and Fiction,
https://doi.org/10.1007/978-3-030-40109-2_7

prefers to drink his vodka martinis "ice-cold" without adding ice. The ice is intercepted by a sieve during pouring. This quite plausible explanation for the *"shaken, not stirred"* is far too profane and truly unworthy of a James Bond. Whatever 007's motivations, there has to be a more elaborate reason for him to require bartenders the world over to make his cocktail in this particular way. So let's have a look for some other possible reasons.

While examining the cocktail conundrum more closely, we came across an article published in the British Medical Journal by six scientists from the Department of Biochemistry at the University of Western Ontario, Canada. The authors entitled their paper *"Shaken, not stirred: bioanalytical study of the antioxidant activities of martinis"*.[2] The amazing result of this study is that the shaken cocktail seems to be healthier than the stirred one. If the cocktail is shaken, the so-called free radicals are better dissolved in the liquid. These free radicals are very aggressive and occur when oxygen molecules are decomposed.[3] Stirred, free radicals remain partially in the drink and can cause damage to the body. Too many free radicals in the body can promote diseases such as cancer, arteriosclerosis, and cataracts. Thanks to his favourite cocktail, 007 seems to be protected from such diseases. The alcohol also contains polyphenols—vegetable tanning agents that are healthy because they can defuse free radicals. So whenever we drink a vodka martini, please let us do so in James Bond style, because the shaken martini destroys twice as many free radicals as the stirred one. James Bond, who leaves nothing to chance, also displays an amazing foresight in the matter of cocktails.

This explanation sounds good and is also "strictly scientific", supported by the Canadian researchers' study.[4] It has only one blemish: if 007 really has his vodka martini shaken because it's healthier, then you could rightly accuse him of being a health fanatic. But neither in Ian Fleming's novels nor in the films is there the slightest evidence for this theory. Surely Bond can't be accused of being obsessed about his health. As we know from the films, he never gives much thought to balancing his diet. This is a man who feasts on beluga caviar, eats oysters by the bucketful, and washes it all down with litres of champagne. He may have a discerning palate, but honestly, he doesn't give a damn about nutrition.[5] It seems unlikely that he would make an exception when it came to cocktails and insist on them being shaken for the good of his health. In any

---

[2] C.C. Trevithick M.M. Chartrand, J. Wahlman, F. Rahman, M. Hirst & J.R. Trevithick, *BMJ* Volume 319, 18 December 1999, pp. 1600–1602. For experts: this journal has an impact factor of 20!

[3] Free radicals also include hydrogen peroxide molecules. These molecules already played a role in the propulsion of the rocket backpack. Free radicals are usually highly reactive chemical substances.

[4] Note, however, that many studies in the field of medicine are not reproducible.

[5] For example, you never see Bond buying his groceries in an organic food store.

case, alcohol is bad for you, whether shaken or stirred. Even the cucumber sandwich offered by Hugo Drax in the movie *Moonraker* is rejected, although it would have been very healthy.

So there must be another reason for shaking the drink. A physicist always tries to approach problems from the fundamental side, i.e., in the case of a vodka martini via the molecular structure of the beverage. A vodka martini is a liquid mixture of relatively large and relatively small molecules. The large molecules are so-called aromatic compounds, which consist of ring-shaped atomic arrangements. Such molecules determine the taste of a drink, as their name suggests. On the other hand, ethanol, i.e., alcohol, is a small, compact molecule. Thus, the cocktail gets its flavor from the large molecules, and its potency from the small ones. But what bearing could this have on 007's drink?

The answer lies in the Brazil nut effect. When a container containing particles of different sizes is shaken, the larger particles rise to the surface.[6] Have you ever opened a box of muesli and wondered why the nuts—the heavier, larger ingredients—are clustered at the top? It's another example of the Brazil nut effect. Figure 7.1 illustrates the effect with large, heavy and small, light balls, where the latter initially all lie on top. After giving the container a good long shake—contrary to intuition—all the large and heavy balls end up on top, while all small and light balls are on the bottom. One reason for this is a packing density effect. Whenever a small gap appears during the shaking process, a small ball can slip down through it. However, there is never such a large gap that a large ball can fall down. That's why only the small balls can move downwards. But if only the small balls can move downwards, then only one direction remains for the large ones: upwards, even if they are much heavier than the small ones. They therefore accumulate at the top of the container.

So there we have our explanation for the traditional request, *"Shaken, not stirred"*. The molecules in a stirred cocktail are distributed evenly throughout the drink, whereas in a shaken cocktail the larger molecules are concentrated at the surface, as shown in Figure 7.2. Crucially, the larger molecules give the cocktail its taste.

Now all we have to do is make it clear who we are thinking about: James Bond! Apparently, 007 already knows when he orders his drink that he will not have time to finish it. On the contrary, as he always rushes from one adventure to the next, he only ever manages one sip of his beloved drink. This one sip should taste as good as possible, however, and the Brazil nut effect

---

[6] More details can be found, for example, in the publications: M.E. Möbius, B.E. Lauderdale, S.R. Nagel, H.M. Jaeger, Nature (London) Vol. 414, 270 (2001) or A.P.J. Breu, H.-M. Ensner, C.A. Kruelle, I. Rehberg, Physical Review Letters Vol. 90, 014302 (2003).

Figure 7.1 Example of the Brazil nut effect. Top left: In a mixture of large, heavy balls (white) and small, light balls (grey), the small balls are initially at the top. Then this mixture is shaken. Top right: Slight shaking causes some of the small balls to drift downwards whenever they encounter a small gap. Bottom left: More and more small balls work their way down into the gaps. Bottom right: After shaking for a certain time, all the large, heavy balls end up at the top and all the small, light ones are at the bottom

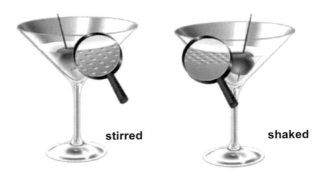

Figure 7.2 In a stirred vodka martini (left), the different molecules are distributed evenly, regardless of size. In a shaken martini, the larger molecules gather on the surface (right). These molecules give the cocktail its flavour

ensures this. The theory presented here thus states that James Bond, despite his exhausting life, is an extreme connoisseur whose palate is equipped with such fine taste buds that he even pays attention to this kind of very small detail. And so the riddle of the shaken vodka martini can be solved satisfactorily, while proving once again that Britain's top spy is a science buff who puts his knowledge of physics into practice even when it comes to apparently trivial matters like ordering a drink.[7]

Naturally, this is only a theory, no less so than the health hypothesis put forward by the Canadian scientists mentioned above. On the other hand, there is proof that the double-O agent is not a health freak, but an extreme connoisseur. In the movie *Never Say Never Again*,[8] 007 is accused by his superior M of being decadent:

M: *"Too many free radicals, that's your problem."*
Bond: *"Free radicals, sir?"*
M: *"Toxins that destroy the body and the brain. Caused by eating too much red meat and white bread, and too many dry martinis."*
Bond: *"Hmmm. Then I shall cut out the white bread, sir."*

---

[7] The reader has hopefully noticed that we are being a little facetious here. Although the Brazil nut effect does actually exist as indicated, it remains up to the reader to decide whether this effect would also work with a vodka martini …

[8] Released in 1983, *Never Say Never Again* is a remake of *Thunderball*. It isn't part of the official canon, but it's a typical Bond adventure with Sean Connery as 007, so it seems justified to cite it here.

8

Epilog: How This Book Came About

007 is in bed with the nuclear physicist Dr. Christmas Jones in a clear posture.

> Bond: *"I was wrong about you."*
> Dr. Jones: *"Yeah. How so?"*
> Bond: *"I thought Christmas only comes once a year!"*
> (Quote from the 19th Bond film *The World Is Not Enough*)

When Piper publishing house approached me almost fifteen years ago and asked if I wanted to write a book about physics in the James Bond films, I never thought it would come to this. Admittedly, I had been giving public lectures on the phenomenal physics knowledge of the secret agent for several years and the adventures of 007 were also among the standard examples of my lectures for physics, engineering, and chemistry students at the TU Dortmund University—but write a book about it?

Two fortunate circumstances had to come together before the project could get started. First of all, I got the support of Prof. Dr. Joachim Stolze, a comrade-in-arms, who spontaneously agreed to work on the book. Then we organized a seminar for the summer semester 2007 with the title *"Shaken, not stirred! We are writing a book about physics in James Bond films"*, announced in the hope that enough participants would come forward to work on the individual topics of the book. Initial doubts as to whether this was a good concept were immediately dispelled. The success was huge, and this is where the second lucky circumstance occurred: a total of 41 students took it upon themselves to thoroughly examine the exploits of top agent Bond during his many adventures.

© The Author(s) 2020
M. Tolan, J. Stolze, *Shaken, Not Stirred!*, Science and Fiction,
https://doi.org/10.1007/978-3-030-40109-2_8

The result became available as a German book in 2008, just before the release of the film *"Quantum of Solace"*. Everyone involved had a lot of fun. If readers enjoy reading the stories half as much as we did calculating the physics involved in all the tricky stunts and gadgets, then this book will have achieved its goal.

This book is not a textbook. The reader is intended to enjoy discovering James Bond's secrets by applying the laws of nature. If you learn something about forces, magnetic fields, lasers, oscillations, and lenses—almost without noticing it—well, all the better! The realization that everyday problems can really be solved with physics is certainly also exciting for many people who have never thought this possible before.[1]

I owe great thanks to Dr. Klaus Stadler and Britta Egetemeier, who worked for the Piper publishing company at the time and encouraged me again and again to go ahead with this project. Without Katharina Wulffius, for many years an editor at Piper, this book project would certainly not have been realized. We would like to thank her for her tireless help, thorough editing, and very pleasant cooperation in producing the book in 2008. I would also like to thank Janine Erdmann from Piper and Florian Feldhaus for their efforts in creating the many graphics.

Without my colleague Prof. Dr. Joachim Stolze and our students Christophe Cauet, Julian Wishahi, Sebastian Jerosch, Dennis Spyra, Kathrin Stich, Nils Uhle, Daniel Pidt, Marco Lafrenz, Björn Wemhöner, Michael Mohr, Björn Bannenberg, Peter Schäfer, Christoph Bruckmann, Florian Feldhaus, Sandra Kuch, Michael Andrzejewski, Philipp Leser, Tobias Brambach, Fabian Clevermann, Ben Wortmann, Marc Daniel Schulz, Helge Rast, Daniel Brenner, Christoph Sahle, Manuela Meyer, Sarah Groß-Bölting, Nils Drescher, Katharina Woroniuk, Frank Hommes, Claudia Zens, Andreas Kim, Anne Hüsecken, Michael Schliwka, Jörn Krones, Sabrina Hennes, Marlene Doert, Steffen Bieder, David Odenthal, Thorsten Brenner, Julia Rimkus, Boris Konrad, and Henrike Enders, this book would certainly never have been written.

After more than 10 years, a complete new edition is now due, which not only includes a complete revision and new arrangement of the chapters, but also many new adventures. In particular, several scenes from the movies with Daniel Craig have been examined, since he had completed only one mission at the time of publication of the 1st edition of the book. Despite several editions, some remaining calculation errors were corrected and the graphics were

---

[1] Admittedly, the everyday problems that James Bond has to endure during his adventures are often very different from those that we have to deal with.

heavily revised. I would like to thank Anne Stadler, Maike Hannen, Martin Kulik, and in particular Charlyne Bieniek from Piper Verlag for their great support, and again Janine Erdmann, Florian Feldhaus, and Julia Peduzzi for the production of the many new and revised graphics. I would also like to thank Springer for publishing the English version of the book. Here I am indebted to Angela Lahee for her great support for this project and to Stephen Lyle for converting deepL English into real English by an extremely careful proofreading.

Metin Tolan, December 2019

Printed in the United States
By Bookmasters

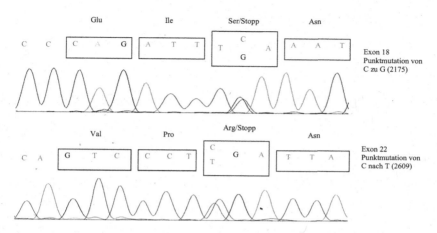

Abbildung 4.3 Mutationen im Proenteropeptidase-Gen bei Patient 2. In Exon 18 liegt eine Punktmutation von C zu G vor (Position 2175 auf der cDNA-Sequenz). Dies hat ein Stoppcodon zur Folge. In Exon 22 liegt gleichfalls eine Punktmutation von C zu T vor, was ebenfalls ein Stoppcodon erzeugt (Position 2609 auf der cDNA-Sequenz)

Lebenslauf von Cornelius Bück

Persönliche Daten und Bildung:

Geburtsdatum: 28.09.1972
Geburtsort: Esslingen am Neckar (D), aufgewachsen in Tübingen (D)
Gymnasium: 1983–1992 Geschwister-Scholl-Schule in Tübingen (D)
Zivildienst: 1992–1993 Behindertenpflege in Freiburg im Breisgau (D)
Studium: 1993–1995 zunächst Psychologie und Physik an der Eberhard-Karls-Universität Tübingen (D)
        1995–2003 anschliessend Humanmedizin an der Ludwig-Maximilians-Universität München (D)

Berufserfahrung: Klinische Tätigkeit (CH):

2019-heute leitender Oberarzt Privatklinik Wyss Münchenbuchsee

2017–2019 Oberarzt Forensik Psychiatrische Dienste Graubünden (PDGR)

2015–2016 PUK Zürich, Allgemeinpsychiatrie 2013–2015 Eigene Praxis Zürich (20 % Pensum)

Facharzttitel «Psychiatrie und Psychotherapie» am 29.11.2012, Bezirksärztekammer Südwürttemberg (D)

2011–2015 Oberarzt UPD Bern, Allgemein- und Sozialpsychiatrie 2010–2015 Kurse, Supervisionen und Selbsterfahrungen am C.

G. Jung-Institut Küsnacht/Zürich

2010 Assistenzarzt Psychiatrische Dienste Thurgau, Allgemeinpsychiatrie

© Der/die Herausgeber bzw. der/die Autor(en), exklusiv lizenziert an Springer Fachmedien Wiesbaden GmbH, ein Teil von Springer Nature 2022
C. Bück, *Sequenz und Struktur des Pro-Enteropeptidase-Gens als Basis für Mutationsanalysen bei angeborenem Enteropeptidasemangel*,
https://doi.org/10.1007/978-3-658-40167-2

Klinische Tätigkeit (D):

2008–2009 Neurologie Fremdjahr; Assistenzarzt am Verbundkrankenhaus Bernkastel/Wittlich

Erlangung des Doktortitels an der LMU München (Magna cum laude) am 24.04.2008

2007 Suchttherapie; Assistenzarzt am Zentrum für Psychiatrie (ZfP) Südwürttemberg in Bad Schussenried

2003–2006 Allgemeinpsychiatrie; Assistenzarzt an der Psychiatrischen Klinik der TU München; Approbation als Arzt am 01.10.2004 (Regierung von Oberbayern)

Literaturverzeichnis

1. Eggermont E, Molla A M, Tygat G, Rutgeerts L (1971) Distribution of enterokinase activity in the human intestine. Acta Gastroenterol Belg 34:655–662
2. Follet G F, Macdonald T H (1976) Intestinal enterokinase deficiency. Acta Paediatr Scand 65:653–655
3. Ghishan F K, Lee P C, Lebenthal E, Johnson P, Bradley C, Greene H (1983) Isolated congenital enterokinase deficiency. Recents findings and review of literature. Gastroenterology 85:727–731
4. Goddon R (1977) Revue critique. Le déficit congénital en entérokinase. Pédiatrie 32:297–301
5. Green J R, Bender S W, Posselt H G, Lentze M (1984) Primary intestinal enteropeptidase deficiency. J Pediatr Gastroenterol Nutr 3:630–633
6. Hadorn H-B, Tarlow M, Lloyd J, Wolff O (1969) Intestinal Enterokinase Deficiency. Lancet 1: 812–813
7. Hadorn H-B (1998) Universitäre Pädiatrie im Wandel. Was wird morgen anders sein? Emeritierungsvorlesung vom 9. Juli 1998. Unveröffentlicht
8. Haworth J, Gourley B, Hadorn H-B, Sumida C (1971) Malabsorption and growth failure due to intestinal enterokinase deficiency. J Pediatr 78:481–490
9. Haworth J, Hadorn H-B, Gourley B, Prasad A, Troesch V (1975) Intestinal enterokinase deficiency. Arch Dis Child 50:277–282
10. Holzinger A, Maier E M, Bück C, Mayerhofer P U, Kappler M, Haworth J C, Moroz S P, Hadorn H-B, Sadler J E, Roscher A A (2002) Mutations in the proenteropeptidase gene are the molecular cause of congenital enteropeptidase-deficiency. Am J Hum Genet 70:20–25
11. Ioannou P A, Amemiya C T, Garnes J, Kroisel P M, Shizuya H, Chen C, Batzer M A, de Jong P J (1994) A new bacteriophage P1-derived vector for the propagation of large human DNA fragments. Nat Genet 6:84–89
12. Kitamoto Y, Veile R, Donis-Keller H, Sadler J E (1995) cDNA sequence and chromosomal localisation of human enterokinase, the proteolytic activator of trypsinogen. Biochemnistry 34:4562–4568
13. Kitamoto Y, Yuan X, Wu Q, McCourt D, Sadler J E (1994) Enterokinase, the initiator of intestinal digestion, is a mosaic protease composed of a distinct assortment of domains. Proc. Natl. Acad. Sci 91: 7588–7592

14. LaVallie E R, Rehemtulla A, Racie L A, DiBlasio E, Ferenz C, Grant K L, Light A, McCoy J M (1993) Cloning and functional expressing of cDNA encoding thr catalytic subunit of bovine enterokinase. J Biol Chem 268:23311–23317

15. Lebenthal E, Antonowicz I, Schwachmann H (1976) Enterokinase and Trypsin activities in pancreatic insufficiency and disease of small intestine. Gastroenterology 70:508–512

16. Lentze M, Green J, Strechi E, Nüssle D, Hermier M (1982) Intestinal enteropeptidase deficiency assiciated with exocrine pancreatic insufficiency. The Lancet 1: 504

17. Lewin, B (1997) Genes VI. Oxford University Press

18. Mann, N S, Mann S K (1994) Enterokinase. Proc Biol Med 206:114–118

19. Marshall G, Mitchel J D, Tobias V, Messina I M (1989) Arrhythmogenic right ventricular dysplasia in a child with congenital enteropeptidase deficiency and hypogammaglobulinaemia. Aust Paediatr J 25:106–108

20. Pardou S, Cardranel S, Rodesch P, Eggermont E, Loeb H (1975) Déficience en entérokinase. Pédiatrie 30:544

21. Pawlow I P (1956) Die gegenwärtige Vereinigung der Hauptseiten der Medizin im Experiment; gezeigt am Beispiel der Verdauung. Akademie-Verlag Berlin II/2:432–450

22. Pawlow I P (1956) 24. Vorlesung: Die Galle als Aktivator des Steapsins – Die Bildung der Galle – Erreger des Gallenaustritts in den Verdauungskanal – Der Darmsaft und die Methodik seiner Gewinnung – Der Brunnersche Abschnitt des Zwölffingerdarms. Akademie-Verlag Berlin V:149–156

23. Pawlow I P (1956) 25. Vorlesung: Die Erreger der Darmsaftabsonderung – Die chemische Zusammensetzung des Darmsaftes – Die Erreger der Kinasereaktion. Akademie-Verlag Berlin V:157–162

24. Pawlow I P (1956) 26.Vorlesung: Die Absonderung des festen Anteils des Darmsaftes – Die Absonderung der Kinase – Die Innervation der Kinasereaktion. Akademie- Verlag Berlin V:162–167

25. Polonovski C, Laplane, R, Alison F, Navarro J (1970) Pseudo-Déficit en Trypsinogene par déficit congenital en Entérokinase. Arch Fr Pédiatri 27:677–688

26. Schepowalnikow N P (1900) Die Physiologie des Darmsaftes. Jahresbericht über die Fortschritte der Thier-Chemie oder der physiologischen und pathologischen Chemie 29:378–380

27. Sleisinger F, Fortran H (1998) Gastrointestinal and Liver Disease.

28. Tarlow M, Hadorn H-B, Arthurton M, Lloyd J (1970) Intestinal Enterokinase Deficiency. Arch Dis Child 45:651–655

29. Watson J D et al (1993) Rekombinierte DNA. Spektrum Akademischer Verlag.

30. Yuan X, Zheng X, Lu D, Rubin D, Pung C, Sadler J E (1998) Structure of murine enterokinase (enteropeptidase) and expression in small intestine during development. Am J Physiol 274:G342–G349

31. Zamolodchikova T, Sokolova E, Lu D, Sadler J E (2000) Activation of recombinant proenteropetidase by duodenase. FEBS Lett 466: 295–299

Publikationen

1. Holzinger A, Maier E M, **Bück C**, Mayerhofer P U, Kappler M, Haworth J C, Moroz S P, Hadorn H-B, Sadler J E, Roscher A A (2002) Mutations in the proenteropeptidase gene are the molecular cause of congenital enteropeptidase deficiency. Am J Hum Genet 70:20–25

2. Holzinger A, Maier E M, **Bück C**, Mayerhofer P U, Haworth J C, Moroz S, Sadler J E, Roscher A A, Hadorn H-B (2000) Mutations in the Proenteropeptidase Gene are the Primary Cause of Congenital Enteropeptidase Deficiency. Journal of Pediatric Gastroenterology and Nutrition, Abstr. 498; Vol. 31; Supplement 2

3. Holzinger A, Maier E M, **Bück C**, Mayerhofer P U, Haworth J C, Moroz S, Sadler J E, Roscher A A, Hadorn H-B (2001) Mutations in the Proenteropeptidase Gene are the Primary Cause of Congenital Enteropeptidase Deficiency (Abstract). Eur. J. Pediatr. 160:3; p R24

4. Holzinger A, Maier E M, **Bück C**, Mayerhofer P U, Haworth J C, Moroz S, Sadler J E, Roscher A A, Hadorn H-B (2000) Erstmalige Identifizierung von Mutationen im Enteropetidase-Gen bei angeborenem Enteropeptidasemangel (Vortrag). Jahrestagung der Süddeutschen Gesellschaft Für Kinderheilkunde und Jugendmedizin in München

5. Holzinger A, Maier E M, **Bück C**, Mayerhofer P U, Haworth J C, Moroz S, Sadler J E, Roscher A A, Hadorn H-B (2000) Mutations in the Proenteropeptidase Gene are the Primary Cause of Congenital Enteropeptidase Deficiency (Abstract). 1st World Congress of Pediatric Gastroenterology, Hepatology and Nutrition in Boston

6. Holzinger A, **Bück C**, Maier E M, Mayerhofer P U, Roscher A A, Sadler J E, Hadorn H-B (1999) Genbankeintrag vom 23. Juni 1999. http://www.ncbi.nlm.nih.gov/. Accession Nos Y19124–Y19143.

Printed in the United States
by Baker & Taylor Publisher Services